D1500275

Springer Tracts in Advanced Robotics
Volume 10

Editors: Bruno Siciliano · Oussama Khatib · Frans Groen

Springer

Berlin
Heidelberg
New York
Hong Kong
London
Milan
Paris
Tokyo

Engineering ONLINE LIBRARY

springeronline.com

B. Siciliano · A. De Luca · C. Melchiorri · G. Casalino (Eds.)

Advances in Control of Articulated and Mobile Robots

With 124 Figures and 15 Tables

Springer

Professor Bruno Siciliano, Dipartimento di Informatica e Sistemistica, Università degli Studi di Napoli Federico II, Via Claudio 21, 80125 Napoli, Italy, email: siciliano@unina.it

Professor Oussama Khatib, Robotics Laboratory, Department of Computer Science, Stanford University, Stanford, CA 94305-9010, USA, email: khatib@cs.stanford.edu

Professor Frans Groen, Department of Computer Science, Universiteit van Amsterdam, Kruislaan 403, 1098 SJ Amsterdam, The Netherlands, email: groen@science.uva.nl

STAR (Springer Tracts in Advanced Robotics) has been promoted under the auspices of EURON (European Robotics Research Network)

Editors

Prof. Bruno Siciliano
Dip. di Informatica e Sistemistica
Università di Napoli Federico II
Via Claudio 21
80125 Napoli, Italy
siciliano@unina.it

Prof. Alessandro De Luca
Dip. di Informatica e Sistemistica
Università di Roma "La Sapienza"
Via Eudossiana 18
00184 Roma, Italy
deluca@dis.uniroma1.it

Prof. Claudio Melchiorri
Dip. di Elettronica, Informatica
e Sistemistica
Università di Bologna
Via Risorgimento 2
40136 Bologna, Italy
cmelchiorri@deis.unibo.it

Prof. Giuseppe Casalino
Dip. di Informatica, Sistemistica
e Telematica
Università di Genova
Via all'Opera Pia 13
16145 Genova, Italy
pino@dist.unige.it

ISSN 1610-7438

ISBN 3-540-20783-X   Springer-Verlag Berlin Heidelberg New York

Cataloging-in-Publication Data applied for
A catalog record for this book is available from the Library of Congress.
Bibliographic information published by Die Deutsche Bibliothek
Die Deutsche Bibliothek lists this publication in the Deutsche Nationalbibliografie; detailed bibliographic data is available in the Internet at <http://dnb.ddb.de>.

Springer-Verlag is a part of Springer Science+Business Media

springeronline.com

© Springer-Verlag Berlin Heidelberg 2004
Printed in Germany

Typesetting: Digital data supplied by author.
Data-conversion and production: PTP-Berlin Protago-TeX-Production GmbH, Berlin
Cover-Design: design & production GmbH, Heidelberg
Printed on acid-free paper   62/3020 Yu - 5 4 3 2 1 0

Foreword

At the dawn of the new millennium, robotics is undergoing a major transformation in scope and dimension. From a largely dominant industrial focus, robotics is rapidly expanding into the challenges of unstructured environments. Interacting with, assisting, serving, and exploring with humans, the emerging robots will increasingly touch people and their lives.

The goal of the new series of *Springer Tracts in Advanced Robotics (STAR)* is to bring, in a timely fashion, the latest advances and developments in robotics on the basis of their significance and quality. It is our hope that the greater dissemination of research developments will stimulate more exchanges and collaborations among the research community and contribute to further advancement of this rapidly growing field.

Advances in Control of Articulated and Mobile Robots edited by Bruno Siciliano, Alessandro De Luca, Claudio Melchiorri, and Giuseppe Casalino provides a unique collection of a sizable segment of the robotics research in Italy. It reports on contributions from ten academic institutions brought together within MISTRAL, an Italian project on robotics research.

This ten-chapter volume covers important research areas ranging from planning, control, and actuation of articulated mechanisms to sensing, perception, navigation, and real-time control architectures of mobile robots. The focus is on fundamental issues related to robots subjected to nonholonomic constraints, time delays, actuator saturation, or joint friction. The work also addresses other key issues concerned with the localization and mapping in unknown or partially known environments, the presence of moving objects, the use of multiple sensors, and the integration of mobility and manipulation.

The thorough discussion, rigorous treatment, and wide span of the work unfolding in these areas reveal the significant advances in the theoretical foundation and technology basis of the robotics field. MISTRAL culminates with this important reference to the world robotics community on the current developments and new directions undertaken by this project's Italian robotics team!

Stanford, California
November 2003

Oussama Khatib
STAR Editor

Preface

Since the development of robotics for industrial and manufacturing applications in structured environments, research in the field has been gradually seeking at providing robotic systems with enhanced autonomy for operation in unstructured environments. Significant examples include cooperating and assisting robots, haptic interfaces for virtual reality and remote operation in hostile environments, mobile robots and autonomous agent teams. The challenge presented by such themes demands advanced control techniques and architectures to perform robotic tasks such as manipulation, interaction, teleoperation, locomotion and cooperation.

This monograph stems from the research project MISTRAL (Methodologies and Integration of Subsystems and Technologies for Anthropic Robotics and Locomotion), funded in 2001–2002 by the Italian Ministry for Education, University and Research (MIUR), involving a significant portion of the national academic robot control community; namely, the research groups at: University of Bologna, University of Genoa, Polytechnical University of Marche, Polytechnic of Milan, University of Naples, University of Pisa, University of Rome "La Sapienza", University of Rome "Tor Vergata", Third University of Rome, Polytechnic of Turin. A complete description of the project is available at the web site *http://www-lar.deis.unibo.it/mistral*.

The aim of this monograph is to provide an updated source of information on the state of the art in advanced control of articulated and mobile robots, along with a taste of significance and impact of new research in the field. A number of relevant problems have been selected dealing with enhanced actuation, motion planning and control functions for articulated robots, as well as of sensory and autonomous decision capabilities for mobile robots.

The material has been organized as follows. The first two chapters are devoted to tutorial/survey presentations on two critical issues when controlling a robotic system: planning motion in the presence of differential constraints, and copying with time delay in remote operation, respectively. The remaining contents have been ordered in a progressive way; the next four chapters deal with control of articulated robots, whereas the final four chapters are focused on planning, localization and servoing of mobile robots. A reading track along the various contributions of the ten chapters of the volume is outlined in the following.

The volume starts with a comprehensive tutorial by *De Luca et al.* on motion planning for a class of *robotic systems subject to nonholonomic differential constraints*. Of special concern is the problem of planning point-to-point motion for systems subject to non-integrable first and second-order differential constraints. The solutions outlined for both non-flat nonholonomic kinematic systems and flat underactuated dynamic systems demonstrate the generality of the approach.

Teleoperation has historically been one of the pioneering areas in robotics. The key problem from a control viewpoint has been to cope with time delay. The chapter by *Arcara and Melchiorri* presents an extensive survey of the most adopted techniques for telemanipulation. Control schemes are critically compared in terms of suitable criteria, and one type of passive controller is analyzed in detail for performance enhancement purposes.

As outlined above, the issue of performance plays a crucial role in robot control. The chapter by *Morabito et al.* concentrates on a specific phenomenon which may deteriorate performance in a robot manipulator undergoing actuator torque saturation. An effective *anti-windup control* law is proposed which is remarkably based on simple and intuitive parameter tuning.

The following two chapters are devoted to the problem of modelling and compensation of *nonlinear friction* in robot joint actuators, yet another effect which must be properly taken into account when designing advanced control systems. The chapter by *Ferretti, Magnani and Rocco* demonstrates how the use of high-resolution encoders allows an accurate analysis of the dynamic behavior of friction forces in the so-called presliding regime, and especially in the presence of hysteresis loops.

On the other hand, the treatment of nonlinear friction in the chapter by *Bona, Indri and Smaldone* is framed into the context of *rapid prototyping* of model-based robot controllers. General issues related to both hardware and software architectures are critically surveyed with the goal of achieving fast and systematic interaction between the algorithmic design phase and the experimental testing.

The use of visual sensors is argued to have high impact for operation in unstructured environments, especially if the robot is visually servoed in a closed-loop control fashion. The problem of *visual tracking* of 3D objects is treated in the chapter by *Caccavale et al.*, where a combined Extended Kalman Filter/Binary Space Partition tree technique is developed to achieve real-time estimation of the position and orientation of moving objects of known geometry using a fixed stereo camera system.

The remaining four chapters deal with issues concerning mobile robots. The development of a *real-time control architecture* for a prototype of differentially-driven wheeled mobile robot is discussed in the chapter by *Bellini et al.*. The solution resorts to RTLinux operating system which seems to gain increasing popularity within the research community; the software architecture includes low level motor feedback, high level trajectory loops, and communication protocols through an Ethernet radio link.

The chapter by *Casalino and Turetta* addresses the problem of coordinating the manoeuvring of a nonholonomic vehicle with the motion of a supported manipulation system, composed either by a single arm or by two arms. Kinematic redundancy is suitable exploited to optimize a number of constraints according to a systematic approach which ensures modularity and scalability within the overall *vehicle-manipulator robotic system*.

Sensory data fusion is covered in the chapter by *Bonci et al.*, where different methods and algorithms are introduced for the accurate localization of mobile robots on a given map, by integration of odometric, gyroscope, sonar and video camera measures using a Kalman filtering approach. On the other hand, different probabilistic methods are employed for the exploration of unknown environments.

The volume ends with the chapter by *Bicchi et al.* which considers three main problems arising in the navigation of autonomous vehicles in partially or totally unknown environments; namely, *map building, localization, and motion servoing.* The result is a generalization of SLAM, which allows the localization and mapping problems to be cast in a unified framework with the control problem.

The monograph is addressed to postgraduate students, researchers, scientists and scholars who wish to broaden and strengthen their knowledge in control of robotic systems.

Besides thanking all the Authors for their valuable contributions to this monograph, we wish to extend our appreciation to all the participants to the MISTRAL project who have produced significant research results during the latest two years. Warmest thanks are also for Thomas Ditzinger at Springer-Verlag in Heidelberg. A final word of thanks goes to Costanzo Manes for the pictorial illustration below.

Italy   *Bruno Siciliano*
October 2003   *Alessandro De Luca*
                                                                                    Claudio Melchiorri
                                                                                      Giuseppe Casalino

Contents

**On the Problem of Simultaneous Localization, Map Building,
and Servoing of Autonomous Vehicles** 223
Antonio Bicchi, Federico Lorussi, Pierpaolo Murrieri, Vincenzo Scordio

Planning Motions for Robotic Systems Subject to Differential Constraints

Alessandro De Luca, Giuseppe Oriolo, Marilena Vendittelli, and Stefano Iannitti

Dipartimento di Informatica e Sistemistica
Università di Roma "La Sapienza"
Via Eudossiana 18, 00184 Roma, Italy
<*deluca,oriolo,vendittelli*>*@dis.uniroma1.it, stefano.iannitti@asi.it*
http://labrob.ing.uniroma1.it

Abstract. We consider the problem of planning point-to-point motion for general robotic systems subject to non-integrable differential constraints. The constraints may be of first order (on velocities) or of second order (on accelerations). Various nonlinear control techniques, including nilpotent approximations, iterative steering, and dynamic feedback linearization, are illustrated with the aid of four case studies: the plate-ball manipulation system, the general two-trailer mobile robot, a two-link robot with flexible forearm, and a planar robot with two passive joints. The first two case studies are non-flat nonholonomic kinematic systems, while the last two are flat underactuated dynamic systems.

1 Introduction

In this chapter, we consider the problem of planning admissible transfer motions for robotic systems that are subject to nonintegrable differential constraints. Such constraints on the motion of a robot may arise from the system mechanical structure (perfect rolling of wheels, conservation of angular momentum) as well as from a reduced control capability (passive degrees of freedom).

The differential constraints can be classified as first-order (i.e., involving velocities) or second-order (involving accelerations). Whenever these constraints are not integrable (or, nonholonomic), the robot may reach a generic point of its state space through suitable maneuvers that are compatible with the constraints. The planning problem consists in generating algorithmically these maneuvers, possibly with a given transfer time. In particular, for first-order kinematic systems we should find a sequence of velocity input commands driving from a given initial configuration to a desired configuration. For second-order dynamic systems, the problem is to find a sequence of force/torque input commands that allow a desired state to be reached from a given initial state, both typically equilibria. As will become clear later in the chapter, the dynamic problem can be often solved by finding a sequence of acceleration inputs on a feedback equivalent second-order (purely kinematic) system.

In order to solve these planning problems, various model transformation techniques can be used, mostly arising from the field of nonlinear control theory. In particular, the possibility of transforming the robot model by means of nonlinear

B. Siciliano et al. (Eds.): Advances in Control of Articulated and Mobile Robots, STAR 10, pp. 1–38, 2004.
© Springer-Verlag Berlin Heidelberg 2004

feedback laws and change of coordinates into a nilpotent system [25], a chained-form system [32], or even a linear controllable system [23] has lead to the definition of powerful planning algorithms.

In particular, we may be able to transform the original nonlinear system into a set of decoupled chains of input-output integrators by means of a dynamic feedback linearizing law [23]. This is possible whenever the state and the input of the system can be expressed algebraically in terms of some output (vector) function and of its derivatives up to a finite order, a strong property called flatness [19]. If a flat output is known for a robot subject to differential constraints, the planning problem can be considered as essentially solved (except for possible singularity issues). This is the case of a large class of wheeled mobile robots (which are subject to nonholonomic first-order kinematic constraints), see e.g. [18,35,47,36], and of robot manipulators including joint elasticity (which are subject to nonholonomic second-order dynamic constraints), see [14].

Therefore, one can basically use the presence or not of the flatness property in order to assess the difficulty of the planning problem in the presence of differential constraints. Necessary and sufficient conditions of flatness are available for nonlinear driftless systems with two inputs [42]. For example, all nonholonomic first-order kinematic systems with two inputs that can be transformed in chained form are flat (and vice versa). However, even when a system is known to be flat but the flat output is not provided, the search for such an output may be not trivial (as in the case of a car towing only one off-hooked trailer [43] or of the bi-steerable vehicle [44]). In addition, assuming that a flat output has been found, it should not be overlooked that singularities may occur in the associated transformations, affecting thus the global validity of the planning algorithm. Unfortunately, there exist no necessary *and* sufficient conditions for flatness (equivalently, for dynamic feedback linearization) in the case of general nonlinear systems with drift. For underactuated robots, which are subject to nonholonomic second-order constraints, the problem is emphasized by the higher complexity of the associated dynamic models.

In any case, the violation of the necessary conditions for flatness given in [42] indicates that the planning problem is not an easy one: this is what happens in the two kinematic case studies presented in this chapter. Moreover, even if some underactuated robots are known to be flat (see, e.g., [1,17]), a deeper analysis of specific planning solutions and of singularities are of interest in the dynamic case. This is the subject of the two other case studies presented later on.

Indeed, there exist other algorithmic approaches to planning motion for systems subject to differential constraints. We just mention here the recently introduced kinematic reduction method for dynamic models of underactuated robots [9]. Based on the concept of kinematic controllability, it is possible in some cases to backup a dynamic motion planning problem into a sequence of elementary velocity commands along so-called decoupling vector fields (see, e.g., [1] for the application to a planar 3R robot with the last passive joint).

The chapter is organized as follows. In Section 2, we review the modeling steps and the properties of kinematic systems with first-order differential constraints, of dynamic systems with first-order differential constraints, and of dynamic systems with second-order differential constraints. In doing so, we also set up the terminology. In the remaining two sections, we address the planning problem for a number of robotic examples that have not been treated extensively in the literature. In particular, two non-flat nonholonomic first-order kinematic systems are considered in Section 3: the plate-ball manipulation system and the general two-trailer wheeled mobile robot. In Section 4, two flat underactuated second-order dynamic systems are presented: a two-link robot with flexible forearm and a planar robot with two passive joints. The presented planning algorithms are based on the use of general mathematical tools investigated by our research group: nilpotent approximations, iterative steering, and dynamic feedback linearization. These concepts will be briefly summarized along the presentation. All case studies include numerical simulation results of the planning of either configuration-to-configuration transfer tasks (in kinematic systems) or of rest-to-rest state transfers (in dynamic systems). We also address robustness issues of the iterative planner for the plate-ball system (Section 3.1) and present a simple planner for the flexible robot in the case of multiple deformation modes (Section 4.1), for which a flat output is not known.

2 Modeling

Let $q = (q_1, \ldots, q_n)$ be a set of n configuration variables of the robotic system. For simplicity, we shall assume that the configuration space of the robot is $I\!R^n$. Moreover, if there were some *holonomic* (geometric) constraints involving the system coordinates, we suppose that such constraints have been already eliminated by suitably reducing the dimension of the configuration space. Therefore, q are generalized coordinates in the Lagrangian sense.

2.1 Kinematic Systems with First-Order Differential Constraints

Assume that a set of $n - m \geq 1$ scalar differential constraints of the form

$$a_i^T(q)\dot{q} = 0 \qquad i = 1, \ldots, n - m, \tag{1}$$

are imposed on the robot motion. The rows $a_i^T(q)$ can be reorganized into a matrix, so that the constraints are rewritten in the compact form

$$A^T(q)\dot{q} = 0. \tag{2}$$

These homogeneous constraints are called *Pfaffian*, being linear in the generalized velocities $\dot{q}$. They may arise from several physical phenomena, most notably the perfect rolling of robot wheels on the ground, the rolling of the fingers of a dextrous robot hand in contact with an object, the conservation of zero angular momentum in free-flying space robots. Under the hypothesis that the columns of matrix A are linearly independent at every q, it follows from (2) that, at a given configuration q, the set of admissible generalized velocities $\dot{q}$ is restricted to a subspace of dimension $m < n$ of $I\!R^n$.

We are interested in the case where the set of constraints (2) is *completely non-holonomic*[1], i.e., when none of the single constraints (1) nor any combination of them through functions $\gamma_i(q)$ is integrable to a holonomic constraint $h(q) = 0$. To check this, nonlinear controllability techniques can be used. The following construction characterizes all feasible instantaneous motions allowed by the differential constraints (2). Define an $(n \times m)$ matrix $G(q)$ whose columns $g_i(q)$, $i = 1, \ldots, m$, are independent vector fields at any q and such that

$$\mathcal{R}\left(G(q)\right) = \mathcal{N}\left(A^T(q)\right), \tag{3}$$

or $A^T(q)G(q) = 0$, for all $q \in \mathbb{R}^n$. Therefore, we can generate all instantaneous feasible velocities $\dot{q}$ as

$$\dot{q} = G(q)v = \sum_{i=1}^{m} g_i(q)v_i. \tag{4}$$

Different choices can be made for defining a matrix $G(q)$ that satisfies (3). Typically, a good choice should be 'physically' motivated, in the sense that the weights v_i, $i = 1, \ldots, m$, represent identifiable (pseudo-)velocities in the robotic system. By assuming that $v \in \mathbb{R}^m$ is the control input, we refer to (4) as the *first-order kinematic model* of the robotic system subject to the first-order differential constraints (2). This model is in the form of a nonlinear driftless control system. By Frobenius theorem on integrability of differential forms, the complete nonholonomy of (2) is equivalent to the *accessibility* of the whole configuration space $\mathbb{R}^n$ of control system (4).

We note also that, in spite of the 'kinematic' terminology, the differential constraints (2), and thus the control system (4), may contain dynamic parameters (i.e., related to the robot mass and inertia). For example, this happens when (2) stems from conservation of generalized momenta.

2.2   Dynamic Systems with First-Order Differential Constraints

One can also take into account the dynamics of a robotic system in the presence of the first-order differential constraints (2). In this case, the model explicitly contains Lagrange multipliers $\lambda \in \mathbb{R}^{n-m}$, representing the generalized constraint forces. The dynamic model in the Lagrangian form is [20, p. 45]

$$B(q)\ddot{q} + n(q, \dot{q}) = A(q)\lambda + S(q)\tau \tag{5}$$
$$A^T(q)\dot{q} = 0, \tag{6}$$

with

$$n(q, \dot{q}) = \dot{B}(q)\dot{q} - \frac{1}{2}\frac{\partial}{\partial q}\left(\dot{q}^T B(q)\dot{q}\right) + \frac{\partial U(q)}{\partial q},$$

---

[1] While each of the scalar differential constraints (1) may not be integrable, a subset of $p < n - m$ or the entire set of $n - m$ differential constraints may still be integrable. In the former case we have *partially nonholonomic* constraints, while in the latter we obtain $n - m$ *holonomic* constraints. In both cases, a reduction of the dimension of the configuration space is induced.

and where $B(q)$ is the $(n \times n)$ symmetric positive definite inertia matrix (so that $\frac{1}{2}\dot{q}^T B(q)\dot{q}$ is the system kinetic energy), $U = U(q)$ is the system potential energy (due, e.g., to gravity or elasticity), $\tau \in I\!\!R^m$ is the force/torque control input, and $S(q)$ is an $(n \times m)$ input matrix which is assumed to be full (column) rank.

Under suitable hypotheses, it is possible to eliminate the Lagrange multipliers λ and to reduce accordingly the set of dynamic equations [10]. Since $G^T(q)A(q) = 0$, premultiplying (5) by $G^T(q)$ leads to a reduced set of m second-order differential equations

$$G^T(q)\,(B(q)\ddot{q} + n(q,\dot{q})) = G^T(q)S(q)u. \tag{7}$$

We can merge the kinematic model (4) (i.e., all generalized velocities $\dot{q}$ satisfying (6)) into (7) so as to obtain

$$\dot{q} = G(q)v$$
$$M(q)\dot{v} + m(q,v) = G^T(q)S(q)\tau, \tag{8}$$

with

$$M(q) = G^T(q)B(q)G(q) > 0$$
$$m(q,v) = G^T(q)B(q)\dot{G}(q)v + G^T(q)\,n(q,G(q)v)$$

and where the vector of pseudo-velocities $v \in I\!\!R^m$ is now part of the system state. Note that the dimension of the state (q,v) has been reduced to $n + m$.

Assuming that 'enough control' is available, or

$$\det\,\left(G^T(q)S(q)\right) \neq 0,$$

we can use a nonlinear static state feedback in order to further simplify (8). Define the control input τ as

$$\tau = \left(G^T(q)S(q)\right)^{-1}\left(M(q)a + m(q,v)\right), \tag{9}$$

where $a \in I\!\!R^m$ is the vector of pseudo-accelerations. The resulting system is

$$\dot{q} = G(q)v$$
$$\dot{v} = a. \tag{10}$$

It is clear that the feedback law (9) leads to model equations that are simply an *extension* (i.e., obtained by the addition of one integrator on each of the m scalar inputs) of the first-order kinematic model (4). We shall thus refer to (10) as the *second-order kinematic model* of the robotic system subject to the first-order differential constraints (2). This model is in the form of a nonlinear control system, with the pseudo-acceleration vector a as input, and contains now a drift term of kinematic nature.

2.3   Dynamic Systems with Second-Order Differential Constraints

A different situation arises when there are no first-order differential constraints of the type (1) but the dynamic system is *underactuated*, i.e., it has less control inputs than degrees of freedom. Let $p \in I\!\!R^n$ be the generalized coordinates (the change of notation will be clear in a moment) and $\tau \in I\!\!R^m$ the available control forces/torques, with $m < n$.

The Lagrangian dynamic equations are of the form

$$B_p(p)\ddot{p} + n_p(p,\dot{p}) = S(p)\tau \tag{11}$$

with a similar notation as in (5) and the same assumption that the $(n \times m)$ input matrix $S(p)$ is full column rank. This model covers various interesting situations, such as for example: a robot with $n - m$ unactuated/failed (in any case, passive) joints; a robot including transmission (joint) elasticity, for which $n = 2m$ and $p = (\theta, \phi)$, being $\theta \in I\!\!R^m$ and $\phi \in I\!\!R^m$, respectively, the positional coordinates of the motors and of the driven links; a robot having flexible links, where $p = (\theta, \delta)$, being $\theta \in I\!\!R^m$ the positions of the motors at the link bases and $\delta \in I\!\!R^{n_e}$ the generalized coordinates describing the deflection of the links, with $n = m + n_e$.

Equation (11) can be elaborated in order to have a set of $n - m$ intrinsic second-order dynamic constraints appear more explicitly. Let $S_l(p)$ be a left inverse of the input matrix $S(p)$ (e.g., the pseudoinverse $S^{\#} = (S^T S)^{-1} S^T$) and $S^{\perp}(p)$ an $((n - m) \times n)$ matrix whose rows annihilate matrix $S(p)$, or $S^{\perp}(p)S(p) = 0$ for any $p \in I\!\!R^n$. Such two matrices can always be chosen so that a coordinate transformation $q = Q(p)$ exists whose Jacobian is (at least locally) nonsingular and equals

$$J_Q(p) = \frac{\partial Q(p)}{\partial p} = \begin{bmatrix} S_l(p) \\ S^{\perp}(p) \end{bmatrix}^{-T}.$$

From (11), one has

$$B_p(p) \begin{bmatrix} S_l(p) \\ S^{\perp}(p) \end{bmatrix}^T \left(\ddot{q} - \frac{d}{dt}\left(\frac{\partial Q(p)}{\partial p} \right)\dot{p} \right) + n_p(p,\dot{p}) = S(p)\tau.$$

This leads to new dynamic equations in the form

$$B(q)\ddot{q} + n(q,\dot{q}) = \begin{bmatrix} S_l(p) \\ S^{\perp}(p) \end{bmatrix} S(p)\tau = \begin{bmatrix} \tau \\ 0 \end{bmatrix}, \tag{12}$$

with

$$B(q) = J_Q^{-T}(p)B_p(p)J_Q^{-1}(p)\Big|_{p=Q^{-1}(q)}$$

$$n(q,\dot{q}) = J_Q^{-T}(p)\left(n_p(p,\dot{p}) - B_p(p)J_Q^{-1}(p)\dot{J}_Q(p)\dot{p} \right)\Big|_{\dot{p}=J_Q^{-1}(p)\dot{q},\, p=Q^{-1}(q)}$$

At this stage, the new coordinates q can be partitioned as $q = (q_a, q_u)$, with actuated coordinates $q_a \in I\!\!R^m$ and unactuated coordinates $q_u \in I\!\!R^{n-m}$. Accordingly, the dynamic model (12) becomes

$$\begin{bmatrix} B_{aa}(q) & B_{ua}^T(q) \\ B_{ua}(q) & B_{uu}(q) \end{bmatrix} \begin{bmatrix} \ddot{q}_a \\ \ddot{q}_u \end{bmatrix} + \begin{bmatrix} n_a(q,\dot{q}) \\ n_u(q,\dot{q}) \end{bmatrix} = \begin{bmatrix} \tau \\ 0 \end{bmatrix}, \tag{13}$$

with blocks of appropriate dimensions. In particular, the last $n - m \geq 1$ equations in (13) can be rewritten separately as

$$A_u^T(q)\ddot{q} + n_u(q,\dot{q}) = [\, B_{ua}(q) \quad B_{uu}(q) \,] \begin{bmatrix} \ddot{q}_a \\ \ddot{q}_u \end{bmatrix} + c_u(q,\dot{q}) + e_u(q) = 0, \quad (14)$$

where the vector $n_u(q,\dot{q})$ has been separated into the Coriolis and centrifugal terms $c_u(q,\dot{q})$ and the potential terms $e_u(q) = (\partial U/\partial q_u)^T$. Note that matrix $A_u^T(q)$ has always full row rank, equal to $n - m$, at any q.

Equation (14) represents a set of $n - m$ *second-order (dynamic) differential constraints* that have to be satisfied by any admissible robot trajectory. The above constraints are linear in the acceleration $\ddot{q}$. At a given state $(q,\dot{q})$, the set of admissible generalized accelerations $\ddot{q}$ is restricted to a linear subspace of dimension m. The complete non-integrability of the set of constraints (14), in the sense of [37], indicates that the underactuated robot can be considered as a mechanical system with *second-order nonholonomic constraints*. As a particular case, it is immediate to see that, whenever $e_u(q) \not\equiv 0$, the constraints $A_u^T(q)\ddot{q} + n_u(q,\dot{q}) = 0$ cannot be obtained from the differentiation of Pfaffian constraints $A_u^T(q)\dot{q} = c$ (a state constraint that would imply a reduction of the state space).

A convenient normal form for the underactuated dynamics (13) is obtained by using again nonlinear static state feedback. Solving (14) for $\ddot{q}_u$ and substituting in the first set of (13), one can verify that the (globally defined) control law

$$\tau = \Big(B_{aa}(q) - B_{ua}^T(q) B_{uu}^{-1}(q) B_{ua}(q) \Big) a + n_a(q,\dot{q}) - B_{ua}^T(q) B_{uu}^{-1}(q)\, n_u(q,\dot{q}) (15)$$

gives

$$\begin{aligned} \ddot{q}_a &= a \\ B_{uu}(q)\, \ddot{q}_u &= -B_{ua}(q)\, a - n_u(q,\dot{q}), \end{aligned} \qquad (16)$$

with the actuated coordinates now directly controlled by the generalized acceleration input $a \in \mathbb{R}^m$. The control (15) is commonly referred to as a *partial feedback linearization* law. In the control system (16), it is clear that the inertial coupling term $B_{ua}(q)$ between actuated and passive coordinates plays a decisive role in the *controllability* properties of the system.

3 Planning for Non-Flat Kinematic Systems

With the aid of two case studies, we shall now illustrate a general technique which achieves asymptotic (in a sense to be clarified below) planning for non-flat kinematic systems subject to differential constraints. In particular, we will consider the plate-ball manipulation system and a wheeled mobile robot, the so-called general two-trailer system. The reader is referred to [48] and [38] for details.

3.1 The Plate-Ball Manipulation System

Rolling manipulation has recently attracted the interest of robotic researchers as a convenient way to achieve dexterity with a relatively simple mechanical design (see [33,6,30] and the references therein). In fact, the nonholonomic nature of rolling contacts between rigid bodies can guarantee the controllability of the manipulation system (hand+manipulated object) with a reduced number of actuators. More in general, this is another example of the minimalistic trend in the field of robotics, aimed at designing devices of reduced complexity for performing complex tasks.

The archetype of rolling manipulation is the plate-ball system [31,27,24,8]: the ball (the manipulated object) can be brought to any contact configuration by maneuvering the upper plate (the first finger), while the lower plate (the second finger) is fixed. Despite its mechanical simplicity, the planning and control problems for this device already raise challenging theoretical issues. In fact, in addition to the well-known limitations coming from its nonholonomic nature, the plate-ball system is neither flat nor nilpotentizable; therefore the classical techniques for nonholonomic motion planning cannot be applied.

To this date, the planning problem has been solved through the symbolic algorithm of [27] and the numerical algorithm of [30]. These techniques, however, are heavily dependent on the specific geometry of rolling surfaces and are not amenable to any kind of generalization to systems of different nature. Our objective is instead to show that asymptotic, robust planning for the plate-ball mechanism can be simply achieved through iterative application of an appropriate open-loop control law designed for the nilpotent approximation of the system. This paradigm, based on the theoretical results in [29], is general and applicable to a wide variety of non-flat systems.

Kinematic model Consider the system shown in Fig. 1, consisting of a spheric ball of radius ρ rolling between two horizontal plates. The lower plate is fixed, while the upper is actuated and can translate horizontally. Denote by u and v the coordinates (latitude and longitude, respectively) of the contact point on the sphere, by x, y the Cartesian coordinates of the contact point on the lower plane, and by ψ the angle between the x axis and the plane of the meridian through the contact point. We assume $-\pi/2 < u < \pi/2$ and $-\pi < v < \pi$, so that the contact point belongs always to the same coordinate patch for the sphere.

The manipulation system is completely described by the kinematics of contact between the sphere and the lower plate [31]. Assume that w_x and w_y, the Cartesian components of the translational velocity of the sphere, are directly controlled[2]. In view of the nilpotent approximation procedure, it is convenient to triangularize the system through the input transformation

$$\begin{bmatrix} w_x \\ w_y \end{bmatrix} = \begin{bmatrix} -\sin\psi\cos u & \cos\psi \\ -\cos\psi\cos u & \sin\psi \end{bmatrix} \begin{bmatrix} w_1 \\ w_2 \end{bmatrix}.$$

---

[2] Recall that the translational velocity of the sphere is half the translational velocity of the upper plane.

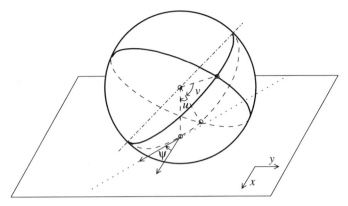

Fig. 1. The plate-ball system. The upper plate is not shown in the figure for the sake of clarity.

This transformation is always defined, except for $u = \pm\pi/2$ which is however outside our coordinate patch. We obtain

$$
\begin{bmatrix} \dot{u} \\ \dot{v} \\ \dot{\psi} \\ \dot{x} \\ \dot{y} \end{bmatrix} = \begin{bmatrix} 0 \\ 1/\rho \\ -\sin u/\rho \\ -\sin\psi\cos u \\ -\cos\psi\cos u \end{bmatrix} w_1 + \begin{bmatrix} 1/\rho \\ 0 \\ 0 \\ \cos\psi \\ -\sin\psi \end{bmatrix} w_2. \tag{17}
$$

Nilpotent approximation Nilpotent approximations [21,4] of nonlinear systems are high-order local approximations that are useful when tangent linearization does not retain controllability, as in nonholonomic systems. In particular, the computation of approximate steering controls for the original system can be performed symbolically, thanks to the closed-form integrability of the nilpotent system, which is polynomial and triangular by construction.

Thanks to the particular structure of our iterative steering strategy (see below), it is sufficient to compute the nilpotent approximation at configurations of the form $\bar{q} = (0, 0, 0, \bar{x}, \bar{y})$. Applying the procedure given in [4] to system (17), one obtains the so-called *privileged* coordinates by the following change of variables

$$
\begin{aligned}
z_1 &= \rho v \\
z_2 &= \rho u \\
z_3 &= \rho^2 \psi \\
z_4 &= -\rho^3 u + \rho^2 (x - \bar{x}) \\
z_5 &= \rho^3 v + \rho^2 (y - \bar{y}).
\end{aligned} \tag{18}
$$

In particular, at $\bar{q}$ one obtains $z = 0$. The transformation is globally valid due to the fact that the degree of nonholonomy is 3 everywhere.

The approximate system is then computed by differentiating eqs. (18) and expanding the input vector fields in Taylor series up to a suitably defined order:

$$
\begin{aligned}
\dot{\hat{z}}_1 &= w_1 \\
\dot{\hat{z}}_2 &= w_2 \\
\dot{\hat{z}}_3 &= -\hat{z}_2 w_1 \\
\dot{\hat{z}}_4 &= -\hat{z}_3 w_1 \\
\dot{\hat{z}}_5 &= \frac{1}{2}\hat{z}_2^2 w_1 - \hat{z}_3 w_2.
\end{aligned}
\tag{19}
$$

The approximation is polynomial and triangular; in particular, the dynamics of $\hat{z}_1$ and $\hat{z}_2$ is exactly the same of z_1 and z_2.

Planning strategy Assume that we wish to transfer the plate-ball system from q^0 to q^d, respectively the initial and desired contact configuration. Without loss of generality, we assume that $q^d = (0,0,0,0,0)$; this can always be achieved by properly defining the reference frames on the sphere and the lower plane.

Our objective is to devise an asymptotic planning strategy; if possible, we would also like robustness with respect to the presence of model perturbations (e.g., on the sphere radius ρ). To this end, it is necessary to embed some form of feedback into the planning method. A natural way to realize this is represented by the iterative steering (IS) paradigm [29]. The essential tool of this method is a contractive open-loop control law, which can steer the system closer to the desired state q^d in a finite time. If such a control is Hölder-continuous with respect to the desired reconfiguration, its *iterated* application (i.e., from the state reached at the end of the previous iteration), guarantees exponential convergence of the state to q^d. The overall input is a time-varying law which depends on a sampled feedback action. A certain degree of robustness is also achieved: a class of non-persistent perturbations is rejected, and the error is ultimately bounded in the presence of persistent perturbations.

To comply with the IS paradigm outlined above, we must design an open-loop control that steers system (17) from q^0 to a point closer in norm to $q^d = (0,0,0,0,0)$. Since the plate-ball manipulation system is controllable [27], such an open-loop control certainly exists. However, the necessary and sufficient conditions for flatness [19] are not satisfied; equivalently, the system cannot be put in chained form, as already noticed in [30]. Therefore, we cannot use conventional techniques for generating the required open-loop control. We therefore settle for an approximate (but symbolic) solution; this is on the other hand consistent with the IS framework, which only requires the error to contract at each iteration.

Our open-loop controller requires two phases:

I. Drive the first three variables u, v and ψ to zero. This amounts to steering the ball to the desired contact configuration regardless of the variables x and y, i.e., of the Cartesian position of the contact point. Denote by $q^I = (0,0,0,x^I,y^I)$ the contact configuration at the end of this phase.

II. Bring x and y closer to x^d and y^d (in norm), while guaranteeing that u, v and ψ return to their desired zero value.

Since the first three equations of (17) can be easily transformed in chained form, phase I can be performed in a finite time T_1 by choosing one of many available steering controls for such systems (see [26]). However, the latter should comply with the Hölder-continuity requirement with respect to the desired reconfiguration; relevant examples are given in [29].

For the second phase, a possible choice is to perform a cyclic motion of period T_2 on u, v and ψ, giving final values $x(T_1 + T_2) = x^{II}$ and $y(T_1 + T_2) = y^{II}$ closer to zero than $x(T_1) = x^I$, $y(T_1) = y^I$. To design a control law that produces such a motion, we shall exploit the nilpotent approximation of the plate-ball system.

Consider the nilpotent dynamics (19) computed at the approximation point q^I. The synthesis of a control law that transfers in time T_2 the state $\hat{z}$ from $z^I = 0$ to z^{II} (respectively, the images of q^I and $q^{II} = (0, 0, 0, x^{II}, y^{II})$, computed through eqs. (18)) can be done as follows. Choose the open-loop control inputs as

$$w_1 = a_1 \cos \omega t + a_2 \cos 4\omega t \tag{20}$$

$$w_2 = a_3 \cos 2\omega t, \tag{21}$$

with $a_1, a_2, a_3 \in I\!R$ and $\omega = 2\pi/T_2$.

Integration of Eqs. (19) shows that in order to obtain $z_4(T_2) = z_4^{II}$ and $z_5(T_2) = z_5^{II}$, coefficients a_1 and a_2 in (20), (21) must be chosen as

$$a_1 = \sqrt{\frac{z_4^{II}}{k_1 a_3}} \qquad a_2 = \frac{z_5^{II}}{k_2 a_3^2}, \tag{22}$$

having set $k_1 = -T_2^3/32\pi^2$ and $k_2 = T_2^3/128\pi^2$. The value of a_3 is immaterial as long as (i) $a_3 \neq 0$ when $z_4^{II} \neq 0$ or $z_5^{II} \neq 0$, and (ii) $\text{sign}(a_3) = -\text{sign}(z_4^{II})$. Therefore, denoting by $\| \cdot \|$ the Euclidean norm, we can let

$$a_3 = -\text{sign}(z_4^{II}) \cdot \left\| \begin{bmatrix} z_4^{II} \\ z_5^{II} \end{bmatrix} \right\|^{1/2r} \qquad r > 1, \tag{23}$$

This choice guarantees for a_1, a_2 and a_3 the Hölder-continuity property required by the IS paradigm.

The other condition to be met by our two-phase open-loop control is contraction of the *original* system (17) from q^0 to q^{II} in spite of (i) the drift of x and y to x^I and y^I due to the first phase (ii) the approximation error[3] induced on x and y by the use of the nilpotent dynamics (19) for computing a steering control. It may be shown (see [39] for details) that contraction is guaranteed provided that a suitable definition of norm is used (to take care of the first-phase drift) and a sufficiently small contraction is required from z^I to z^{II} (to reduce the approximation error within admissible bounds).

---

[3] Note that u, v and ψ return to zero under the proposed open-loop inputs, as verified by integration of the first three equations of the original system (17). Thus, the open-loop controls (20), (21) are exactly cyclic in u, v and ψ.

Iterative steering We now clarify the use of the proposed open-loop controller within the IS framework to achieve an asymptotic planner.

Starting from the initial contact configuration, apply the open-loop control of phase I for the required time T_1. Using the values x^I, y^I at the end of this phase, the desired z_4^{II} and z_5^{II} are generated as

$$z_4^{II} = \beta_1 z_4^d \qquad z_5^{II} = \beta_2 z_5^d, \tag{24}$$

where $\beta_1 < 1$, $\beta_2 < 1$ are the chosen contraction rates and z_4^d, z_5^d are the images of $x^d = 0$, $y^d = 0$ as given by (18), in which $\bar{x} = x^I$, $\bar{y} = y^I$. At this point, Eqs. (22), (23) are used to compute coefficients a_i, and the phase II open-loop controls (20), (21) are applied to system (17). After $T_1 + T_2$ seconds from the initial time, the system state is sampled and the two-phase control procedure is repeated. In particular, the values of z_4^{II} and z_5^{II} are updated at each iteration using (24) (with constant β_1, β_2). In fact, since transformation (18) depends on the approximation point, the same is true for z_4^d, z_5^d. Note also that:

- Since all the conditions of the IS paradigm are satisfied for β_1, β_2 sufficiently close to 1, it is guaranteed that the manipulation system state q exponentially converges to the desired contact configuration q^d.
- In the absence of perturbations, there is no need to repeat phase I after the first iteration.
- In perturbed conditions, it is necessary to analyze the structure of the perturbation itself. If certain requisites (see [29, Th. 2]) are met, the perturbation will be rejected on the simple basis of the stable behavior of the nominal system.

We may therefore conclude that we have obtained *asymptotic planning* for the plate-ball system, on the basis of the fact that the system variables q converge to the desired configuration q^d. In practice, one can stop the iterations when q is within a prespecified distance of the destination; using the properties of IS, it is also possible to predict the number of iterations needed to achieve a certain error tolerance. The robustness with respect to perturbations is a consequence of the intrinsic sampled feedback nature of the proposed planner.

Simulation results Two simulations are now presented to show the effectiveness of the proposed planner: in the first, perfect knowledge of the system is assumed (*nominal* case), while in the second we have included a perturbation on the ball radius ρ (*perturbed* case).

In the first simulation, we assume that the radius $\rho = 1$ is exactly known and phase I has already been executed. The initial and desired configurations are $q^0 = (0, 0, 0, 0.5, 0.5)$ and $q^d = (0, 0, 0, 0, 0)$, respectively. In each iteration, the open-loop control (20), (21) is applied with $T_2 = 1$ sec, $r = 1.5$ in eq. (23), and contraction rates $\beta_1 = \beta_2 = 0.4$ in (24).

Figure 2 illustrates the exponential convergence of the state variables along the iterations. The Cartesian path of the contact point is shown in Fig. 3: note how the path of the single iterations 'shrinks' with time.

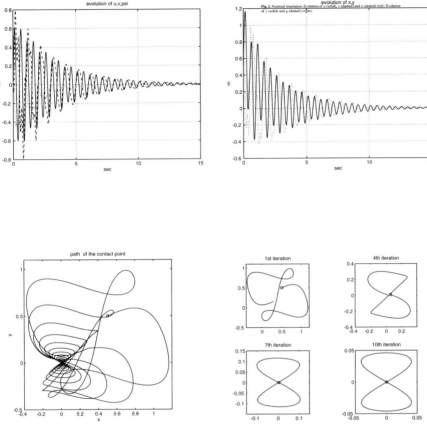

Fig. 3. Nominal simulation: Cartesian path of the contact point; the small circle indicates q^0 (*left*). Cartesian paths of the contact point during the 1st, 4th, 7th and 10th iterations; the small circle indicates the starting configuration of each iteration; notice the different scale in the plots (*right*).

In the second simulation, q^0, q^d as well as the planner parameters are the same of the previous simulation, but a 10% perturbation on the value of the ball radius has been introduced; only its nominal value $\rho = 1$ is known and used for computing the control law. The theoretical framework of the IS paradigm guarantees that this kind of perturbation will be rejected by the iterative steering scheme. Figure 4 confirms that exponential convergence is preserved despite the perturbation — only at a slightly smaller rate. The Cartesian path of the contact point is very similar to the nominal case, as shown in Fig. 5, although the paths in the single iterations are deformed.

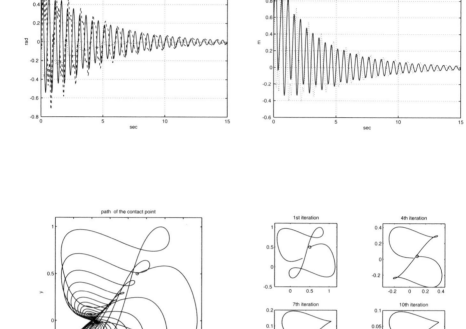

Fig. 5. Perturbed simulation: Cartesian path of the contact point; the small circle indicates q^0 (*left*). Cartesian paths of the contact point during the 1st, 4th, 7th and 10th iterations; the small circle indicates the starting configuration of each iteration (*right*).

3.2   The General Two-Trailer Wheeled Mobile Robot

Another interesting example of non-nilpotentizable, non-flat nonholonomic robot is the general N-trailer system, i.e., a vehicle in which N off-hooked trailers are attached to a tractor. It is well known that this system is non-flat if $N \geq 2$ (see [19] for a proof in the case $N = 2$). The problem of controlling this system has only been addressed so far in [28], where it is shown that at particular configurations the system can be approximated by a chained form. However, the latter are not dense in the state space, so that the method does not apply for generic configurations.

Below, we consider a particular case, i.e., the general two-trailer system, proving that asymptotic planning can be achieved by means of the iterative steering technique based on the nilpotent approximation of the system.

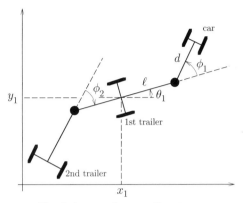

Fig. 6. A general two-trailer system.

Kinematic model Consider the system shown in Fig. 6, consisting of a car towing two identical trailers, each hooked at a distance d from the preceding wheel axle (*off-hooking*). The distance between the hooking point and the wheel axle midpoint of each trailer is ℓ. For simplicity, we assume $d = 1$ and $\ell = 1$. However, a similar analysis can be developed for the case $d \neq \ell$.

With an eye to the nilpotent approximation procedure, it is convenient to choose an appropriate set of generalized coordinates and control inputs. In particular, let $q = (x_1, y_1, \theta_1, \phi_1, \phi_2)$, where x_1, y_1 are the Cartesian coordinates of the first trailer reference point, θ_1 is the first trailer orientation with respect to the x axis, and ϕ_1, ϕ_2 are the angles formed by the car and the first trailer respectively with the first and the second trailer. Also, denote by v_1 and ω_1 the driving and steering velocities of the first trailer, which are related to v_0 and ω_0, the driving and steering velocities of the car (the actual inputs) by the input transformation

$$v_0 = v_1 \cos\phi_1 + \omega_1 \sin\phi_1$$
$$\omega_0 = v_1 \sin\phi_1 - \omega_1 \cos\phi_1,$$

which is always defined. The kinematic model is then obtained as

$$\begin{aligned}
\dot{x}_1 &= \cos\theta_1 \, v_1 \\
\dot{y}_1 &= \sin\theta_1 \, v_1 \\
\dot{\theta}_1 &= \omega_1 \\
\dot{\phi}_1 &= s_1 v_1 - (1 + c_1)\omega_1 \\
\dot{\phi}_2 &= -s_2 v_1 + (1 + c_2)\omega_1,
\end{aligned} \tag{25}$$

having set $s_i = \sin\phi_i$, $c_i = \cos\phi_i$, $s_{ij} = \sin(\phi_i - \phi_j)$ and $c_{ij} = \cos(\phi_i - \phi_j)$ for $i, j = 1, 2$. If $\phi_1 = \pi$ or $\phi_2 = \pi$, the system is clearly not controllable. We consider points of the state space defined as $\mathcal{M} = \mathbb{R}^2 \times S^1 \times (S^1 - \{\pi\})^2$.

Denote by g_1, g_2 the input vector fields of system (25), and consider the first 6 elements of the P. Hall [25] family $g_1, g_2, g_3 = [g_1, g_2]$, $g_4 = [g_1, [g_1, g_2]]$, $g_5 =$

$[g_2, [g_1, g_2]]$, $g_6 = [g_1, [g_1, [g_1, g_2]]]$. Vector fields g_1, g_2, g_3, g_4, g_5 span the tangent space of $\mathcal{M}$ at points such that $\phi_1 \neq \phi_2$ (*regular* points), while g_1, g_2, g_3, g_4, g_6 span the tangent space everywhere, including points such that $\phi_1 = \phi_2$ (*singular* points). Hence, the system is controllable and the degree of nonholonomy is 3 at regular points and 4 at singular points.

Nilpotent approximation In the presence of singular points, homogeneous nilpotent approximations [4] do not provide globally valid representations. However, it has been shown that nonhomogeneous nilpotent forms can be adopted to this end [49]. Applying the procedure therein proposed to system (25), we obtain the following global nilpotent approximation

$$
\begin{aligned}
\dot{\hat{z}}_1 &= u_1 \\
\dot{\hat{z}}_2 &= u_2 \\
\dot{\hat{z}}_3 &= -\hat{z}_2 u_1 \\
\dot{\hat{z}}_4 &= \sum_{j=1}^{2} h_{j4}(\hat{z}_1, \ldots, \hat{z}_3) u_j \\
\dot{\hat{z}}_5 &= \sum_{j=1}^{2} h_{j5}(\hat{z}_1, \ldots, \hat{z}_4) u_j,
\end{aligned}
\tag{26}
$$

in which

$$
\begin{aligned}
h_{j4}(\hat{z}_1, \ldots, \hat{z}_3) &= a_{j4}^2 \hat{z}_1^2 + b_{j4}^2 \hat{z}_1 \hat{z}_2 + c_{j4}^2 \hat{z}_2^2 + d_{j4}^2 \hat{z}_3 \\
h_{15}(\hat{z}_1, \ldots, \hat{z}_4) &= c_{15}^2 \hat{z}_2^2 + a_{15}^3 \hat{z}_1^3 + b_{15}^3 \hat{z}_1 \hat{z}_3 + c_{15}^3 \hat{z}_1^2 \hat{z}_2 \\
&\quad + d_{15}^3 \hat{z}_2 \hat{z}_3 + e_{15}^3 \hat{z}_2^3 + f_{15}^3 \hat{z}_1 \hat{z}_2^2 + g_{15}^3 \hat{z}_4 \\
h_{25}(\hat{z}_1, \ldots, \hat{z}_4) &= d_{25}^2 \hat{z}_3 + a_{25}^3 \hat{z}_1^3 + b_{25}^3 \hat{z}_1 \hat{z}_3 + c_{25}^3 \hat{z}_1^2 \hat{z}_2 \\
&\quad + d_{25}^3 \hat{z}_2 \hat{z}_3 + e_{25}^3 \hat{z}_2^3 + f_{25}^3 \hat{z}_1 \hat{z}_2^2 + g_{25}^3 \hat{z}_4.
\end{aligned}
$$

The coefficients $a_{j4}^2, \ldots, d_{j4}^2$, c_{15}^2, d_{25}^2 and $a_{j5}^3, \ldots, g_{j5}^3$ ($j = 1, 2$) are functions of $\bar{q} = (\bar{x}_1, \ldots, \bar{\phi}_2)$ around which the approximation is computed. Their expressions are quite complicated and are omitted. However, they are not needed for implementing the stabilization method, thanks to the structure of the chosen control input.

Planning strategy In order to transfer the general two-trailer system from an initial point q^0 to a desired point[4] $q^d = (0, 0, 0, \phi_1^d, \phi_2^d)$, we adopt the same strategy of the plate-ball system. To comply with the IS paradigm, we must design an open-loop control that steers system (25) from q^0 to a point closer in norm to q^d.

As before, our open-loop controller requires two phases:

---

[4] This particular choice of the destination does not imply any loss of generality, because it can always be achieved by translating and rotating the world reference frame so as to align with the desired configuration of the first trailer.

I. Drive in finite time the first three variables x_1, y_1 and θ_1 to zero. This amounts to steering the first trailer to its desired configuration regardless of the variables ϕ_1 and ϕ_2, which will converge to generic values ϕ_1^I, ϕ_2^I.

II. Bring ϕ_1 and ϕ_2 closer to ϕ_1^d and ϕ_2^d (in norm), while guaranteeing that x_1, y_1 and θ_1 return to their desired zero value.

Similarly to the plate-ball system, the first three equations of (25) can be easily transformed in chained form (they are, in fact, the equations of a unicycle). Hence, phase I can be easily performed in a finite time T_1 with Hölder-continuous steering controls.

For the second phase, we use again the nilpotent approximation of the system to perform a cyclic motion of period T_2 on x_1, y_1 and θ_1, giving final values $\phi_1(T_1+T_2) = \phi_1^{II}$, $\phi_2(T_1+T_2) = \phi_2^{II}$ closer to zero than $\phi_1(T_1) = \phi_1^I$, $\phi_2(T_1) = \phi_2^I$. We emphasize that, in view of the globality of the representation (26), q^I may be a regular or singular point. The synthesis of a control law that transfers the state of system (26) from $z^I = 0$ (the image of q^I) exactly to z^{II} (the image of q^{II}) is relatively straightforward.

Consider the nilpotent approximation (26) at q^I. Choose the open-loop control inputs as

$$v_1 = a_1 \cos \omega t + a_2 \sin \omega t \tag{27}$$

$$\omega_1 = a_3 \cos 2\omega t, \tag{28}$$

with $a_1, a_2, a_3 \in I\!R, \omega = 2\pi/T$ and T the duration of the control interval. Integration of Eqs. (26) shows that in order to obtain $z_4(T) = z_4^{II}$ and $z_5(T) = z_5^{II}$, parameters a_1 and a_2 in (27–28) can be chosen as

$$a_1 = \sqrt{a_2^2 + \frac{z_4^{II}}{k_1 a_3}} \qquad a_2 = \frac{2\pi}{T} \frac{z_5^{II}}{z_4^{II}} \tag{29}$$

having set $k_1 = -T^3/32\pi^2$ and $k_2 = -T^4/64\pi^3$, and provided that $z_4^{II} \neq 0$. The value of a_3 is immaterial for the steering task, as long as $a_3 \neq 0$ and $\text{sign}(a_3) = -\text{sign}(z_4^{II})$ (so that a_1 is always well defined). In particular, we can let

$$a_3 = -\text{sign}(z_4^{II}) \cdot \left\| \begin{bmatrix} z_4^{II} \\ z_5^{II} \end{bmatrix} \right\|^{1/r} \qquad r > 1. \tag{30}$$

This choice guarantees for a_1, a_2 and a_3 the Hölder-continuity property[5] required by the IS paradigm. In particular:

---

[5] A difficulty with the method so far outlined is that the steering controls (27), (28) are not defined when $z_4^{II} = 0$. On the other hand, Equation (31) gives $z_4^{II} = 0$ if $z_4^d = 0$, i.e., if no reconfiguration is needed for the nilpotent approximation variable z_4. To circumvent this problem, it is relatively easy to work out a more general rule than (31) for generating z_4^{II} and z_5^{II}. In practice, any contraction on the norm of the error $(z_4 - z_4^d \quad z_5 - z_5^d)$ is admissible as long as $z_4^{II} \neq 0$.

- According to (29), a_2 is Hölder-continuous if z_5^{II} converges to zero faster than z_4^{II}. To this end, one simply sets $\beta_1 < \beta_2$ in eq. (31).
- The first coefficient a_1 given by eq. (29) is Hölder-continuous in view of the choice (30) for a_3.

As before, the other condition to be met by our two-phase open-loop control — i.e., contraction of the actual system from q^0 to q^{II} — can be satisfied by suitably choosing the norm and enforcing a sufficiently small contraction on the nilpotent approximation.

Iterative steering Starting from the initial configuration, apply the open-loop control of phase I for the required time T_1. Using the values ϕ_1^I, ϕ_2^I at the end of this phase, the images in privileged coordinates of the final goal values are computed through the change of coordinates between q and z, evaluated on the manifold defined by $x_1 = 0$, $y_1 = 0$, $\theta_1 = 0$:

$$z_4^d = \frac{1}{2} \left(\frac{\phi_2^d - \bar{\phi}_2}{1 + \cos \bar{\phi}_2} - \frac{\phi_1^d - \bar{\phi}_1}{1 + \cos \bar{\phi}_1} \right)$$

$$z_5^d = \frac{1}{2} \left(\frac{\phi_2^d - \bar{\phi}_2}{1 + \cos \bar{\phi}_2} + \frac{\phi_1^d - \bar{\phi}_1}{1 + \cos \bar{\phi}_1} \right)$$

The desired z_4^{II} and z_5^{II} are now generated as

$$z_4^{II} = \beta_1 z_4^d \qquad z_5^{II} = \beta_2 z_5^d, \tag{31}$$

where $\beta_1 < 1$, $\beta_2 < 1$ are the chosen contraction rates.

At this point, Equation (29) is used to compute the parameters a_i, and the phase II open-loop controls (27), (28) are applied to system (25). After $T_1 + T_2$ seconds, the system state is sampled and the procedure is repeated. Since the conditions of the IS paradigm have been satisfied, it is guaranteed that the state q of the general two-trailer system exponentially converges to the desired configuration q^d, and hence asymptotic planning has been achieved. Again, in the absence of perturbations, there is no need to repeat phase I after the first iteration, while in perturbed conditions it is necessary to analyze the structure of the perturbation itself.

Simulation results We present two simulations of the proposed planning strategy. In both cases, it is assumed that phase I has already been executed, so that the first trailer is already at its desired configuration $x_1^d = 0$, $y_1^d = 0$, $\theta_1^d = 0$. Phase II is executed by iterative application of the control inputs (27), (28), with $T = 1$ sec and the coefficients a_i ($i = 1, \ldots, 3$) given by (29), (30), with $r = 4$. The contraction rates in (31) have been chosen as $\beta_1 = 0.6$ and $\beta_2 = 0.7$.

In the first simulation, it is $\phi_1^I = \pi/4$ and $\phi_2^I = -\pi/4$, while the desired values are $\phi_1^d = 0$ and $\phi_2^d = 0$ (a singular configuration). Figure 7 shows the cyclic evolution of x_1, y_1, and θ_1 as well as the trajectory of ϕ_1 and ϕ_2. The motion of the first trailer is shown in Fig. 8 (with different scale on the two axes), which also shows the vehicle configurations at the beginning of phase II, at the end of the first and of the 15-th iteration.

The second simulation starts from $\phi_1^I = \pi/8$, $\phi_2^I = 0$, with the desired configuration given as $\phi_1^d = -\pi/4$, $\phi_2^d = \pi/3$ (a regular point). Figure 9 shows the time evolution of the state variables. Figure 10 reports the Cartesian motion of the first trailer and the configurations of the vehicle at the beginning of phase II, at the end of the first and of the 15-th iteration.

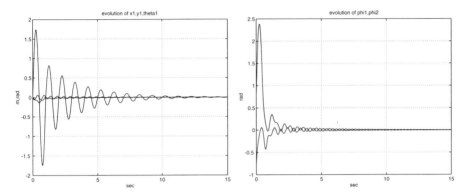

Fig. 7. Simulation 1: Evolution of x_1, y_1 and θ_1 (*left*). Evolution of ϕ_1 and ϕ_2 (*right*).

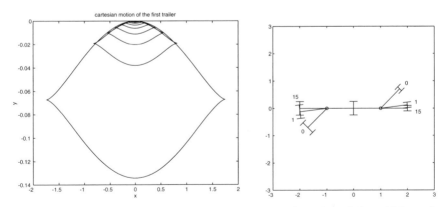

Fig. 8. Simulation 1: Motion of the first trailer (*left*). Configuration of the vehicle at the beginning of phase II (0) after one iteration (1) and after 15 iterations (*right*).

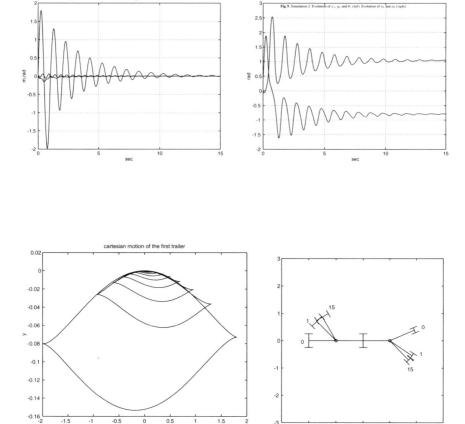

Fig. 10. Simulation 2: Motion of the first trailer (*left*). Configuration of the vehicle at the beginning of phase II (0) after one iteration (1) and after 15 iterations (*right*).

4   Planning for Flat Dynamic Systems

We present two representative case studies of robots with underactuated dynamics for which one can define, under special assumptions, a flat output so that the planning problem can be solved in a relatively easy way. The first system is a two-link planar robot with a flexible forearm. The second system is a 4R planar robot having the last two joints passive and a special hinging condition. For both robots, two actuating inputs are available and motion occurs on a horizontal plane. The reader is referred to [12] and to [22] for details.

4.1   A Two-Link Robot with Flexible Forearm

For a multi-link robot displaying link flexibility, typically encountered in long reach and slender/lightweight arm design [7], the planning of a prescribed reconfiguration between two equilibrium states to be performed in fixed time (*rest-to-rest maneuver*) is a very critical problem. In fact, large and simultaneous motion of the links will induce oscillations that persist beyond the nominal final completion time.

For a single flexible link, characterized by a linear dynamics, there exist model-based techniques, such as input shaping [46] or inverse dynamics trajectory design [3], that allows generating a torque command for rest-to-rest maneuvers. However, these approaches lead only to partial solutions, since motion time is not a design parameter for the input shaping method, while motion completion at the given time is only approximately realized within the non-causal inversion method of [3]. In [11], the problem is tackled by finding the closed-form expression of a (scalar) system output having maximum relative degree, i.e., such that no zeros appear in the transfer function from the input torque to the defined output. As a matter of fact, this output is a flat output for the system and the planning problem is solved by fitting to this output a smooth interpolating polynomial between the start and final rest configurations.

A solution technique for the rest-to-rest problem is not yet available in the case of a general multi-link flexible robot. However, if a flat output vector were found (if one exists), the generalization to the nonlinear setting would be immediate. One such situation occurs in the case of the FLEXARM, a two-link planar robot with a flexible forearm currently available at the Department of Computer Science and Automation of University of Rome Three, provided that flexibility of the forearm is modeled by just one dominant deformation mode.

Dynamic model and partial feedback linearization   The FLEXARM has a first rigid link and a second link that can bend only in the horizontal plane. Due to its mechanical construction, the forearm can be modeled as an Euler-Bernoulli beam (with Young modulus E and cross section inertia I) undergoing small deformations.

Let $\theta_1(t)$ be the angular position of the first link of length ℓ_1 and inertia J_1 (including the first actuator) with respect to the first joint axis. The actuator driving the second link has mass m_{02} and inertia J_{02}. The second flexible link of length ℓ_2 is modeled as a beam of uniform density ρ, mass $m_2 = \rho\ell_2$, and equivalent rigid inertia with respect to the second joint axis $J_2 = m_2\ell_2^2/3$. A payload of mass m_p and inertia J_p can be added at the tip. Define $\theta_2(t)$ as the angular position, with respect to the orientation of the first link, of a line pointing from the second joint axis to the instantaneous center of mass of the flexible forearm (pinned angle).

The transversal bending deformation $w(x,t)$ at a point $x \in [0, \ell_2]$ along the second link is described, in the pinned frame, by separation of space and time as

$$w(x,t) = \sum_{i=1}^{n_e} \phi_i(x)\delta_i(t),$$

where a finite number $n_e \geq 1$ of deformation mode shapes $\phi_i(x)$, with associated deformation coordinates $\delta_i(t)$, have been used. The mode shapes $\phi_i(x)$, for $i =$

$1, \ldots, n_e$, are eigenfunctions (with related angular eigenfrequencies ω_i) associated to the solutions of a fourth-order partial differential equation for $w(x, t)$ subject to suitable geometric/dynamic boundary conditions, and can be computed according to [2,5].

Starting from this analysis, and using the Lagrange-Euler equations of motion, the dynamic model is obtained as

$$B(q)\ddot{q} + n(q, \dot{q}) + Kq = S\tau, \tag{32}$$

with generalized coordinates $q = (\theta, \delta) = (\theta_1, \theta_2, \delta_1, \ldots, \delta_{n_e}) \in I\!\!R^{2+n_e}$. The positive definite inertia matrix $B(q)$ has the structure

$$B(q) = \begin{bmatrix} b_{11}(\theta_2, \delta) & b_{12}(\theta_2, \delta) & b_{13}(\theta_2) & \cdots & b_{1, n_e+2}(\theta_2) \\ & J_{2t} & 0 & \cdots & 0 \\ & & 1 & \ddots & \vdots \\ & \text{symm} & & \ddots & 0 \\ & & & & 1 \end{bmatrix}.$$

For later use, we define $b_\delta = [b_{13} \ldots b_{1, n_e+2}]^T$. The nonlinear Coriolis and centrifugal vector $n(q, \dot{q})$, quadratic in $\dot{q}$, has the structure

$$n(q, \dot{q}) = [n_1(\theta_2, \delta, \dot{\theta}, \dot{\delta}) \quad n_2(\theta_2, \delta, \dot{\theta}_1) \quad n_3(\theta_2, \dot{\theta}_1) \quad \cdots \quad n_{n_e+2}(\theta_2, \dot{\theta}_1)]^T.$$

We define also the subvectors $n_\theta = [n_1 \; n_2]^T$ and $n_\delta = [n_3 \ldots n_{n_e+2}]^T$. Finally, the elasticity matrix K is

$$K = \text{diag} \{0, 0, K_\delta\} = \text{diag} \{0, 0, \omega_1^2, \ldots, \omega_{n_e}^2\},$$

while the input matrix S (transforming the motor torques $\tau = (\tau_1, \tau_2)$ into generalized forces performing work on q) takes on the form

$$S = \begin{bmatrix} I_{2\times 2} & \begin{matrix} 0_{1\times n_e} \\ \Phi'^T(0) \end{matrix} \end{bmatrix}^T = \begin{bmatrix} 1 & 0 & 0 & \cdots & 0 \\ 0 & 1 & \phi_1'(0) & \cdots & \phi_{n_e}'(0) \end{bmatrix}^T.$$

It is apparent that the dynamic system (32) has degree of underactuation equal to n_e. As shown in Section 2.3, it is convenient to apply partial feedback linearization in order to simplify the system equations of an underactuated robot. The dynamic model (32) can be rewritten in block form as

$$\begin{bmatrix} B_{\theta\theta} & B_{\theta\delta} \\ B_{\theta\delta}^T & I \end{bmatrix} \begin{bmatrix} \ddot{\theta} \\ \ddot{\delta} \end{bmatrix} + \begin{bmatrix} n_\theta \\ n_\delta \end{bmatrix} + \begin{bmatrix} 0 \\ K_\delta\delta \end{bmatrix} = \begin{bmatrix} \tau \\ \Phi'(0)\tau_2 \end{bmatrix},$$

partitioned according to the dimensions of θ and δ. Solving for $\ddot{\delta}$ from the second block of equations, substituting into the first, and defining the global nonlinear feedback law for τ as

$$\tau = \begin{bmatrix} 1 & b_\delta^T \Phi'(0) \\ 0 & 1 \end{bmatrix} \left(\begin{bmatrix} b_{11} - b_\delta^T b_\delta & b_{12} \\ b_{12} & J_{2t} \end{bmatrix} \begin{bmatrix} a_1 \\ a_2 \end{bmatrix} + \begin{bmatrix} n_1 - b_\delta^T(n_\delta + K_\delta\delta) \\ n_2 \end{bmatrix} \right), \tag{33}$$

where a_1 and a_2 are new acceleration inputs, leads to an equivalent dynamic model in the form:

$$\ddot{\theta}_1 = a_1$$
$$\ddot{\theta}_2 = a_2 \qquad\qquad\qquad\qquad\qquad\qquad\qquad (34)$$
$$\ddot{\delta} = -b_\delta a_1 - (n_\delta + K_\delta \delta) + \Phi'(0)\,(b_{12}a_1 + J_{2t}a_2 + n_2)\,.$$

For convenience, we detail only the expressions of the terms b_{12}, b_δ, n_2, and n_δ appearing in (34), referring the reader to [12] for the remaining dynamic terms of (32). We have:

$$b_{12} = J_{2t} + h_{n_e+1}\cos\theta_2 - \sum_{i=1}^{n_e} h_i \delta_i \sin\theta_2$$

$$b_{1,i+2} = h_i \cos\theta_2 \qquad i = 1,\dots,n_e$$

$$n_2 = \left(h_{n_e+1}\sin\theta_2 + \sum_{i=1}^{n_e} h_i \delta_i \cos\theta_2\right)\dot{\theta}_1^2$$

$$n_{i+2} = h_i \sin\theta_2\,\dot{\theta}_1^2 \qquad i = 1,\dots,n_e,$$

with $J_{2t} = J_{02} + J_2 + J_p + m_p \ell_2^2$ and the constant coefficients

$$h_i = \left[\rho \int_0^{\ell_2} \phi_i(x)\,dx + m_p \phi_i(\ell_2)\right]\ell_1 \qquad i = 1,\dots,n_e$$

$$h_{n_e+1} = \left[m_2 \frac{\ell_2}{2} + m_p \ell_2\right]\ell_1\,.$$

Planning strategy In a rest-to-rest task, the flexible robot should be moved from an initial configuration $q_i = (\theta_i, 0)$ at time $t_i = 0$ to a final configuration $q_f = (\theta_f, 0)$ at time $t_f = T$, both undeformed and with $\dot{q}(0) = \dot{q}(T) = 0$. We are thus looking for a vector of command torques $\tau(t) = (\tau_1(t), \tau_2(t))$, defined in $t \in [0, T]$, that steers the robot to the goal.

In order to solve this problem, we try to find a two-dimensional output $y = (y_1, y_2)$ having the flatness property. From an operative point of view, one can select an output vector function and then use the dynamic feedback linearization algorithm [23] as a computational tool. In particular, we should be able to differentiate with respect to time the chosen output y a specific number of times until a two-dimensional input appears in a nonsingular way. At some steps of the algorithm, and possibly after a state-dependent change of coordinates in the input space, the addition of integrators on one of the two input channels could be needed, so as to avoid subsequent differentiation of the relative input. This extension process builds up the state of a dynamic compensator. If the total number of output derivatives performed until the input appears equals the number of states of the flexible robot plus the number of added compensator states, then the system is flat, namely it has no zero dynamics and can be transformed via a nonlinear dynamic feedback into two independent chains of integrators from auxiliary inputs to the chosen flat outputs.

We present the application of the dynamic feedback linearization algorithm to the FLEXARM, by taking into account only the first dominant mode of flexible forearm ($n_e = 1$). Equations (34) become

$$\ddot{\theta}_1 = a_1$$
$$\ddot{\theta}_2 = a_2$$
$$\ddot{\delta}_1 = -\omega_1^2 \delta_1 + \phi_1'(0)J_{2t}(a_1 + a_2) + [\,\phi_1'(0)h_1\delta_1 \quad \gamma_1\,]\,R(\theta_2)\begin{bmatrix} \dot{\theta}_1^2 \\ a_1 \end{bmatrix},$$

having set

$$\gamma_1 = \phi_1'(0)J_{2t} - h_1, \qquad R(\theta_2) = \begin{bmatrix} \cos\theta_2 & -\sin\theta_2 \\ \sin\theta_2 & \cos\theta_2 \end{bmatrix}.$$

We choose as candidate flat output

$$y = \begin{bmatrix} y_1 \\ y_2 \end{bmatrix} = \begin{bmatrix} \theta_1 \\ \theta_2 + c_1\delta_1 \end{bmatrix}, \tag{35}$$

where c_1 is a coefficient yet to be defined. Differentiating Eq. (35) twice gives

$$\ddot{y} = \begin{bmatrix} a_1 \\ a_2 + c_1\phi_1'(0)J_{2t}(a_1 + a_2) - c_1\omega_1^2\delta_1 + [\,c_1\phi_1'(0)h_1\delta_1 \quad c_1\gamma_1\,]\,R(\theta_2)\begin{bmatrix} \dot{\theta}_1^2 \\ a_1 \end{bmatrix} \end{bmatrix}.$$

Both acceleration inputs a_1 and a_2 appear at this level, but the total number of output derivatives $(2 + 2 = 4)$ does not yet cover the dimension $2(2 + n_e) = 6$ of the state space. Therefore, in order to make the matrix weighting the inputs in $\ddot{y}$ singular, we can choose the free coefficient c_1 as

$$c_1 = -\frac{1}{\phi_1'(0)J_{2t}}, \tag{36}$$

so that a_2 disappears from the expression of $\ddot{y}_2$. In order to proceed with output differentiation, we need then a dynamic extension on the first input channel (i.e., a_1). In this case, we can directly add two integrators with states denoted by ξ_1 and ξ_2

$$a_1 = \xi_1, \qquad \dot{\xi}_1 = \xi_2, \qquad \dot{\xi}_2 = \alpha_1,$$
$$a_2 = \alpha_2, \tag{37}$$

where $\alpha = (\alpha_1, \alpha_2)$ is the new input. As a result of (36) and (37), $\ddot{y}$ becomes a function of θ_2, $\dot{\theta}_1$, δ_1, and ξ_1 only. The third derivative of the output is still independent from α:

$$y^{[3]} := \frac{d^3y}{dt^3} = \begin{bmatrix} \xi_2 \\ -\xi_2 - c_1\omega_1^2\dot{\delta}_1 + [\,c_1\phi_1'(0)h_1\dot{\delta}_1 \quad 0\,]\,R(\theta_2)\begin{bmatrix} \dot{\theta}_1^2 \\ \xi_1 \end{bmatrix} \\ + [\,c_1\phi_1'(0)h_1\delta_1 \quad c_1\gamma_1\,]\,R(\theta_2)\begin{bmatrix} 2\dot{\theta}_1\xi_1 \\ \xi_2 \end{bmatrix} \\ + \dot{\theta}_2\,[\,c_1\phi_1'(0)h_1\delta_1 \quad c_1\gamma_1\,]\,\frac{dR}{d\theta_2}\begin{bmatrix} \dot{\theta}_1^2 \\ \xi_1 \end{bmatrix} \end{bmatrix}.$$

Thus, through the above expressions of y and its derivatives, a transformation is defined from the original state $(\theta_1, \theta_2, \delta_1, \dot{\theta}_1, \dot{\theta}_2, \dot{\delta}_1)$ and compensator state (ξ_1, ξ_2) to the set of coordinates $(y, \dot{y}, \ddot{y}, y^{[3]}) \in I\!\!R^8$.

By differentiating the output once more, we finally obtain

$$y^{[4]} = A(\theta_2, \delta_1, \dot{\theta}_1, \xi_1)\alpha + f(\theta_2, \delta_1, \dot{\theta}_1, \dot{\theta}_2, \dot{\delta}_1, \xi_1, \xi_2),$$

where the so-called *decoupling matrix* A is

$$A = \begin{bmatrix} 1 & 0 \\ a_{12} & a_{22} \end{bmatrix},$$

with

$$a_{12} = -1 + [\, c_1 \phi'_{p1}(0) h_1 \delta_1 \quad c_1 \gamma_1 \,] R(\theta_2) \begin{bmatrix} 0 \\ 1 \end{bmatrix}$$

$$a_{22} = \omega_1^2 + [\, (c_1 \gamma_1 - \phi'_1(0) h_1) \quad -c_1 \phi'_1(0) h_1 \delta_1 \,] R(\theta_2) \begin{bmatrix} \dot{\theta}_1^2 \\ \xi_1 \end{bmatrix}.$$

The decoupling matrix A is nonsingular iff $a_{22} \neq 0$. Under this assumption (see [12] for a detailed verification), the inversion-based control law defined by the static feedback from the extended (robot + compensator) state

$$\alpha = A^{-1}(\theta_2, \delta_1, \dot{\theta}_1, \xi_1)\left(v - f(\theta_2, \delta_1, \dot{\theta}_1, \dot{\theta}_2, \dot{\delta}_1, \xi_1, \xi_2)\right) \tag{38}$$

transforms the extended dynamic system into a linear controllable one made by two independent chains of four input-output integrators from the auxiliary input $v = (v_1, v_2)$ to the output $y = (y_1, y_2)$, or

$$y^{[4]} = v. \tag{39}$$

Note that (39) represents the whole system, since the total number of output differentiations $(4 + 4 = 8)$ equals the number of states of the flexible robot (6 for $n_e = 1$) plus the number of added compensator states ξ (2 in this case). The dynamic feedback linearizing compensator having as input vector $v = (v_1, v_2)$ and as output the torque vector $\tau = (\tau_1, \tau_2)$ has dimension $\nu = 2$. The complete expression of this compensator is obtained by merging (33), (37) and (38).

Rest-to-rest trajectory generation   Given the initial state at $t = 0$

$$\theta_1(0) = \theta_{1i}, \ \theta_2(0) = \theta_{2i}, \ \delta_1(0) = 0, \ \dot{\theta}_1(0) = \dot{\theta}_2(0) = \dot{\delta}_1(0) = 0$$

and the desired state at $t = T$

$$\theta_1(T) = \theta_{1f}, \ \theta_2(T) = \theta_{2f}, \ \delta_1(T) =, \ \dot{\theta}_1(T) = \dot{\theta}_2(T) = \dot{\delta}_1(T) = 0,$$

by choosing $\xi_1(0) = \xi_2(0) = \xi_1(T) = \xi_2(T) = 0$, one can derive initial and final boundary conditions for the reference output trajectory $y_d(t) = (y_{1d}(t), y_{2d}(t))$ and its derivatives up to the third order. These values can be interpolated by a polynomial trajectory of (at least) 7-th degree (one polynomial for each output) defined for $t \in [0, T]$. Higher-order polynomials can be used in order to achieve a smoother torque profile at the boundaries.

From (38), (39), setting $v = y_d^{[4]}$, we have

$$\alpha_d = A^{-1}(\theta_{2d}, \delta_{1d}, \dot{\theta}_{1d}, \xi_{1d}) \left(y_d^{[4]} - f(\theta_{2d}, \delta_{1d}, \dot{\theta}_{1d}, \dot{\theta}_{2d}, \dot{\delta}_{1d}, \xi_{1d}, \xi_{2d}) \right)$$

where the desired values of the extended state are obtained by inverting the linearizing transformation, in which $y \equiv y_d(t)$ is used at each $t \in [0, T]$.

After substitutions, the nominal rest-to-rest torques are given by

$$\tau_{1d} = \left(b_{11,d} - b_{13,d}^2 \right) \xi_{1d} + b_{12,d}\, \alpha_{2d} + n_{1,d} - b_{13,d}\left(n_{3,d} + \omega_1^2 \delta_{1d} \right)$$
$$+ b_{13,d}\, \phi_1'(0)\left(b_{12,d}\, \xi_{1d} + J_{2t}\, \alpha_{2d} + n_{2,d} \right)$$

$$\tau_{2d} = b_{12,d}\, \xi_{1,d} + J_{2t}\, \alpha_{2d} + n_{2,d},$$

where the added subscript d means that all dynamic model quantities are evaluated along the nominal state trajectory.

Simulation results The FLEXARM is characterized by the following data:

$$
\begin{aligned}
J_1 &= 16.2 \cdot 10^{-4} \text{ kg m}^2 & m_{02} &= 3.118 \text{ kg} \\
\ell_1 &= 0.3 \text{ m} & J_{02} &= 6.35 \cdot 10^{-4} \text{ kg m}^2 \\
EI &= 2.4507 \text{ N m}^2 & \ell_2 &= 0.7 \text{ m} \\
m_p &= J_p = 0 & m_2 &= 1.853 \text{ kg} \\
& & J_2 &= 0.1483 \text{ kg m}^2.
\end{aligned}
\tag{40}
$$

The resulting first eigenfrequency of the forearm is $f_1 = 3.7631$ Hz ($\omega_1 = 2\pi f_1 = 23.6442$ rad/s).

We have considered the following rest-to-rest motion task:

$$\theta_{1i} = \theta_{2i} = 0 \qquad \theta_{1f} = \theta_{2f} = 90° \qquad T = 2 \text{ s.}$$

For each output component in eq. (35), an 11-th order polynomial, with zero symmetric boundary conditions on its derivatives up to the fifth one, has been selected as desired trajectory. This guarantees also boundary continuity, at $t = 0$ and $t = T$, of the rest-to-rest torques and of their first time derivative.

The results in Figs. 11–13 indicate a natural behavior, with bounded deformation in the linearity domain and maximum torques within the actuators capabilities. In particular, two interesting variables for the flexible forearm are the clamped joint angle $\theta_{c2} = \theta_2 + \phi_1'(0)\delta_1$, which is the angular position that can be directly measured by an encoder at the joint, and the tip angle $y_{t2} = \theta_2 + (\phi_1(\ell_2)/\ell_2)\delta_1$, which is the angle between a line pointing at the forearm tip and the x-axis of the pinned frame. In the first half of the motion the clamped angle leads over the second output reference trajectory and the tip lags behind, while the situation is reversed in the second half. The maximum transversal displacement at the forearm tip is about 12 cm.

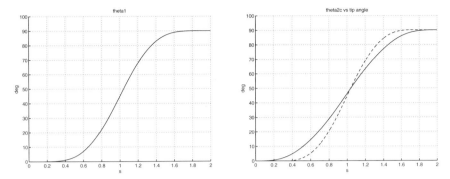

Fig. 11. Motion of first link variable θ_1 (*left*) and of the clamped joint angle θ_{c2} (—) and tip angle y_{t2} (- -) of the flexible forearm (*right*).

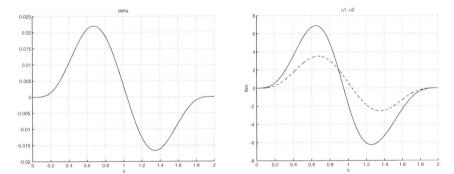

Fig. 12. Evolution of the deformation variable $\delta_1(t)$ of the forearm (*left*) and computed rest-to-rest torques τ_{1d} (—) and τ_{2d} (- -) (*right*).

An extension to the case of multiple modes The above analysis shows that the output (35) (or its natural generalization with $y_2 = \theta_2 + \sum_{i=1}^{n_e} c_i \delta_i$) cannot be flat for the FLEXARM, when $n_e \geq 2$ deformation modes are considered. This is because one can eventually solve (at least locally) for the auxiliary input $\alpha = (\alpha_1, \alpha_2)$ at a differential order that is 'too low' for achieving linearization of the *full state* via dynamic feedback. In fact, the existence of a flat output for $n_e \geq 2$ modes is still an open problem. Nevertheless, it is still possible to design a simple planning algorithm that solves the rest-to-rest motion problem using the following arguments.

The starting point is again the partially feedback linearized model (34), with a generic number of $n_e \geq 2$ flexible modes. For a desired reconfiguration of the robot in a fixed time T, one can split the task in two phases:

I. Move the first link (rigid variable θ_1) to the goal position (with $\dot{\theta}_1 = 0$) in time $T_1 < T$ while keeping the θ_2 variable at its initial rest value. This can

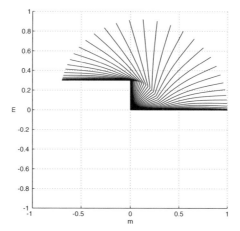

Fig. 13. Stroboscopic view of the FLEXARM (with $n_e = 1$ deformation mode) for a rest-to-rest motion of $T = 2$ s.

be achieved, for instance, using a fifth-order polynomial for the acceleration $a_1(t)$ and setting $a_2(t) = 0$, for $t \in [0, T_1]$. At the end of this first phase, the deformation state of the forearm is denoted as $(\delta^I, \dot{\delta}^I) \neq 0$

II. In the second phase, of duration $T_2 = T - T_1$, we set $a_1(t) = 0$. The dynamics of the flexible robot (with the first link at rest) becomes linear,

$$\ddot{\theta}_2 = a_2 \qquad \ddot{\delta} = -K_\delta \delta + \Phi'(0) J_{2t} a_2,$$

being $n_\delta = 0$ and $n_2 = 0$ for $\dot{\theta}_1 = 0$. This is the dynamics of a one-link flexible arm, so that the method in [11] can be applied for planning the remaining state-to-rest reconfiguration that completes the task. In particular, this is obtained by using a polynomial function $y_{2d}^{II}(t)$ of sufficiently high order that interpolates the proper boundary conditions, at $t = T_1$ and $t = T$, for the scalar output

$$y_2 = \theta_2 + \sum_{i=1}^{n_e} c_i \delta_i \qquad c_i = -\frac{1}{J_{2t} \phi_i'(0)} \prod_{\substack{j=1 \\ j \neq i}}^{n_e} \frac{\omega_j^2}{\omega_j^2 - \omega_i^2},$$

which is in fact a flat output for the forearm subsystem.

Using the same data in (40) for the robot and taking into account $n_e = 3$ flexible modes, we have considered the following rest-to-rest motion task:

$$\theta_{1i} = \theta_{2i} = 0, \qquad \theta_{1f} = \theta_{2f} = 90°, \qquad T = 5 \text{ s}.$$

The switching time between the two phases is $T_1 = 3$ s. In the obtained results of Figs. 14–15, the two motion phases and the larger deformation occurring during the second phase are clearly shown. During phase II, the forearm overshoots and then comes back to the desired position at the prescribed final time. Note that the second torque in phase I keeps the rigid motion component of the second link at rest, while the first torque in phase II keeps the first link at rest.

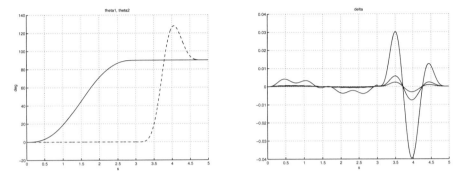

Fig. 14. Variables θ_1 (—) and θ_2 (- -) (*left*) and deformations $\delta_i(t)$ of the forearm (*right*) for a two-phase rest-to-rest motion with $n_e = 3$ flexible modes.

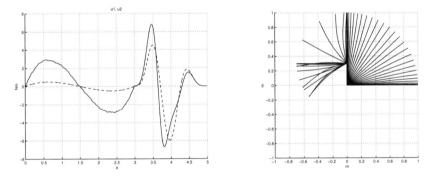

Fig. 15. Computed rest-to-rest torques τ_{1d} (—) and τ_{2d} (- -) (*left*) and stroboscopic view of the FLEXARM (*right*) for a two-phase motion of $T = 5$ s with $n_e = 3$ flexible modes.

4.2   A Planar Robot with Two Passive Joints

Robots with passive joints are purposely designed for saving the cost of actuating each degree of freedom of the mechanical structure or are the result of the occurrence of actuator total failures.

For robots with just one active joint and one or more passive joints, planning of a reconfiguration is in general still an open problem. Existing results are based on the design of stabilizing nonlinear feedback control, thus achieving only an asymptotic planning strategy for reaching the goal configuration (possibly, with an exponential rate of convergence). Examples of this kind can be found in [15] and [13], respectively, for a 2R and a PR robot with only the first (rotational or prismatic) joint actuated.

When there are at least two actuated joints, more planning results are available. A case study that obtained large attention is the planar 3R robot with the last passive joint. The so-called center of percussion[6] (CP) of the third (passive) link has been used for solving rest-to-rest motion problems in [1] and in [16]. In particular, in [1] the planning strategy consists of a sequence of translational and rotational (around the CP point) motions of the third link, while [16] use the fact that the CP position is a flat output for the system. Thanks to partial feedback linearization (see (15)), this result applies whatever is the type of the first two actuated joints. More in general, the CP position of the last link is a flat output for a planar robot with n links having the first $n - 1 > 2$ joints actuated and a last passive rotational joint [17,41] (with or without gravity).

There are few planning results for robots with passive joints having degree of underactuation larger than one (i.e., with at least two passive joints). The only sufficiently general case that has been tackled so far is that of a planar robot with $n \geq 4$ links having the first two joints actuated and the remaining $n - 2$ passive rotational joints. Under a special hinging assumption, namely that each link has the following passive joint axis located at its center of percussion, it has been shown that the CP position of the last link is a flat output for the system [34]. The sequential planning algorithm of [1] has been extended in [45] to this case, while the flatness approach has been detailed in [22]. We summarize here the results of [22] for the case $n = 4$, characterizing also potential dynamic singularities that should be avoided at the planning stage.

Dynamic model and partial feedback linearization We consider the XYRR robot in Fig. 16, a planar structure in the horizontal plane having the two joints proximal to the base can be any combination of prismatic or rotational actuated joints while the two distal joints are passive rotational joints. The degree of underactuation is thus equal to two. It is assumed that the fourth link is hinged exactly at the center of percussion (CP_3) of the third link, which is the same special condition used in [34,45].

The dynamic model of the robot can be derived using the standard Lagrangian formulation. With reference to Fig. 16, and in view of the use of (15) before attacking the planning problem, we shall define the generalized coordinates as $q = (q_a, q_u) = (x, y, q_3, q_4)$, where (x, y) are the Cartesian coordinates of the base of the third link while q_3 and q_4 are the absolute orientations of the last two links with respect to the

---

[6] The center of percussion of a uniform link of length l rotating around one of its end is located at a distance $2l/3$ from the axis of rotation.

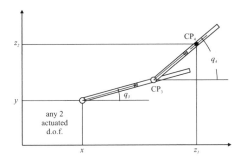

Fig. 16. A general underactuated XYRR robot.

x-axis. Denote by l_i and d_i, respectively, the length of the i-th link and the distance between the i-th joint axis and the i-th link center of mass. Moreover, the distance between the i-th joint axis and the center of percussion CP_i of the i-th link is

$$k_i = \frac{I_i + m_i d_i^2}{m_i d_i}$$

where m_i and I_i are, respectively, the mass and the centroidal moment of inertia of the i-th link. In particular, because of the special hinging condition, we have $k_3 = l_3$.

After partial feedback linearization, the robot dynamic equations take on the form

$$\ddot{x} = a_x$$
$$\ddot{y} = a_y$$
$$l_3 \ddot{q}_3 + \lambda_{34} c_{34} \ddot{q}_4 = s_3 a_x - c_3 a_y - \lambda_{34} s_{34} \dot{q}_4^2 \tag{41}$$
$$l_3 c_{34} \ddot{q}_3 + k_4 \ddot{q}_4 = s_4 a_x - c_4 a_y + l_3 s_{34} \dot{q}_3^2,$$

where we have set for compactness $s_i = \sin q_i$, $c_i = \cos q_i$, $s_{ij} = \sin(q_i - q_j)$, $c_{ij} = \cos(q_i - q_j)$ $(i, j = 3, 4)$ and $\lambda_{34} = m_4 l_3 d_4 / (m_3 d_3 + m_4 l_3)$. Note also that the last two equations have been conveniently scaled here by constant factors.

Planning strategy In a rest-to-rest task, the robot with passive joints should be moved from an initial configuration $q_i = (x_i, y_i, q_{3i}, q_{4i})$ at time $t_i = 0$ to a final configuration $q_f = (x_f, y_f, q_{3f}, q_{4f})$ at time $t_f = T$, with $\dot{q}(0) = \dot{q}(T) = 0$. Starting from the equivalent model (41), we are thus looking for a vector of acceleration input commands $a(t) = (a_x(t), a_y(t))$, defined for $t \in [0, T]$, that steers the robot to the goal.

In order to solve this problem, we use the known flatness property of system (41). As mentioned above, the Cartesian position of CP_4, the center of percussion of the fourth link, is a two-dimensional flat output:

$$\begin{bmatrix} y_1 \\ y_2 \end{bmatrix} = \begin{bmatrix} x + l_3 c_3 + k_4 c_4 \\ y + l_3 s_3 + k_4 s_4 \end{bmatrix}. \tag{42}$$

Following the dynamic linearization algorithm, we need to differentiate six times the output (42) before we can solve (at least locally) for an auxiliary two-dimensional input. In doing so, a dynamic extension by one integrator and an additional static feedback transformation is performed at each step, starting from the second order of differentiation (acceleration level). The dynamic extension on a single channel avoids, as usual, subsequent differentiation of the relative input, whereas the feedback transformation is needed here because the intermediate (2×2) decoupling matrices are singular but have all non-zero entries (see [22] for further details).

The algorithm produces a total addition of four integrators, with states denoted as $\xi_1, \ldots, \xi_4$. We obtain then a dynamic linearizing compensator of dimension $\nu = 4$, with state equations

$$
\begin{aligned}
\dot{\xi}_1 &= \xi_2 \\
\dot{\xi}_2 &= \xi_3 + \dot{q}_4^2 \, \xi_1 \\
\dot{\xi}_3 &= \xi_4 + 2\dot{q}_4^2 \, \xi_2 - \mu \, t_{34} \, \dot{q}_4 \, \xi_1 \\
\dot{\xi}_4 &= u_1 + \phi \, \dot{q}_4 - \psi(\dot{q}_3 - \dot{q}_4)\dot{q}_4
\end{aligned}
\tag{43}
$$

and output equation

$$
\begin{bmatrix} a_x \\ a_y \end{bmatrix} = R(q_3) \begin{bmatrix} \dfrac{1}{c_{34}} \left(\dfrac{k_4 - \lambda_{34} \, c_{34}}{k_4 - \lambda_{34}} \xi_1 + k_4 \, \dot{q}_4^2 \right) + l_3 \, \dot{q}_3^2 \\ u_2 \end{bmatrix},
\tag{44}
$$

where $R(q_3)$ is a planar rotation matrix and we have set

$$
t_{34} = \frac{s_{34}}{c_{34}} \qquad \mu = \frac{\xi_1}{k_4 - \lambda_{34}} + \dot{q}_4^2
$$

$$
\psi = \frac{\mu \xi_1}{c_{34}^2} \qquad \phi = 2\dot{q}_4^3 \, \xi_1 - 3 t_{34} \, \mu \, \xi_2 + 3\dot{q}_4 \, \xi_3 - t_{34} \, \xi_1 \, \dot{\mu}.
$$

The signals u_1 and u_2 are obtained by inverting, at the last step of the algorithm, the expressions of the sixth-order output derivatives in terms of an auxiliary input $v = (v_1, v_2)$:

$$
\begin{aligned}
u_1 &= c_4 v_1 + s_4 v_2 \\
u_2 &= \frac{l_3}{\psi} \left(c_4 v_2 - s_4 v_1 - \dot{q}_4 \, \xi_4 + (\dot{q}_3 - \dot{q}_4)\dot{\psi} - \dot{\phi} + \psi \delta \right),
\end{aligned}
\tag{45}
$$

where

$$
\delta = t_{34} \left(\frac{l_3 + \lambda_{34} \, c_{34}}{l_3(k_4 - \lambda_{34})} \xi_1 + \dot{q}_4^2 \right).
$$

Under the action of the dynamic compensator (43), (45), the robot system has been made equivalent to the linear and controllable form

$$
\begin{bmatrix} y_1^{[6]} \\ y_2^{[6]} \end{bmatrix} = \begin{bmatrix} v_1 \\ v_2 \end{bmatrix},
\tag{46}
$$

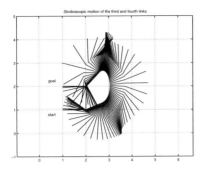
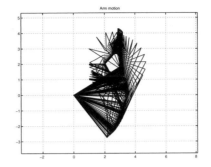

Fig. 17. Stroboscopic motion of the last two links (*left*) and of the whole 4R underactuated robot (*right*).

i.e., two decoupled chains of six integrators each. The total number of output derivatives $(6 + 6 = 12)$ equals the dimension $2n + \nu$ of the extended state space. The linearizing algorithm defines also, in the intermediate steps, a transformation between the robot and compensator states $(q, \dot{q}, \xi) \in I\!R^{12}$ and $(y_1, y_2, \dot{y}_1, \dot{y}_2, \ldots, y_1^{[5]}, y_2^{[5]}) \in I\!R^{12}$. This transformation or, equivalently, the dynamic compensator (43), (45) include however some singularities.

Rest-to-rest trajectory generation Planning a feasible trajectory on the equivalent representation (46) is a smooth interpolation problem for the flat output (y_1, y_2), the position of the center of percussion of the fourth link, with appropriate boundary conditions on the output derivatives up to the fifth order.

The above planning procedure is valid only if the following regularity conditions (compare with the denominators in (43) and (45)) are satisfied throughout the motion:

$$c_{34} \neq 0 \qquad \text{and} \qquad \psi \neq 0.$$

These conditions can be given an interesting physical interpretation. In particular, $c_{34} \neq 0$ means that the third and fourth link should never become orthogonal, while $\psi \neq 0$ holds as long as ξ_1, the acceleration of the CP_4 point *along the fourth link axis*, does not vanish during motion. Besides, since $\xi_1^2 = \ddot{y}_1^2 + \ddot{y}_2^2$, this regularity condition can be checked directly from the planned trajectory for the linearizing outputs. In order to avoid both types of dynamic singularities, the boundary conditions for the compensator state $(\xi_1, \xi_2, \xi_3, \xi_4)$ should be suitably selected at the planning stage.

Simulation results We have considered a 4R underactuated robot with the following (purely kinematic) data for the last two links: $l_3 = k_3 = 1$ m, $l_4 = 1$ m, $k_4 = 2/3$ m, and $\lambda_{34} = 1/3$ m. The first two links have length $l_1 = 3.5$ m and $l_1 = 2.5$ m. The rest-to-rest motion task is defined by

$$q_i = (x_i, y_i, q_{3i}, q_{4i}) = (1, 1, 0, \pi/8) \text{ [m,m,rad,rad]},$$
$$q_f = (x_f, y_f, q_{3f}, q_{4f}) = (1, 2, 0, \pi/4) \text{ [m,m,rad,rad]},$$

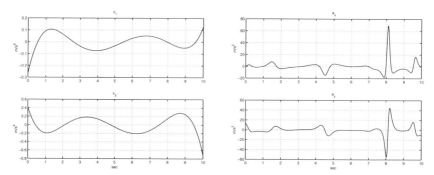

Fig. 18. Evolution of the auxiliary inputs v_1, v_2 (*left*) and of the acceleration inputs a_x, a_y (*right*).

with motion time $T = 10$ s. For each output component in (42), an 11-th order polynomial trajectory has been chosen. The boundary conditions of the associated interpolation problem are evaluated using the initial/final robot state and the initial/final dynamic compensator state. This second set has been chosen symmetrically as

$$(\xi_{1i}, \xi_{2i}, \xi_{3i}, \xi_{4i}) = (\xi_{1f}, \xi_{2f}, \xi_{3f}, \xi_{4f}) = (0.1, 0, 0, 0) \ [\text{m/s}^2, \text{m/s}^3, \text{m/s}^4, \text{m/s}^5].$$

The stroboscopic motion of the last two links and of the whole 4R robot are shown in Fig. 17 (the third and fourth link are represented only until their center of percussion). The two last links undergo a counterclockwise rotation of $360°$, while the first two links never cross a stretched or folded kinematic singularity. The evolution of the auxiliary input $v = (v_1, v_2)$ (namely, the sixth-order time derivatives of the planned output trajectory) and the robot acceleration input $a = (a_x, a_y)$ are shown in Fig. 18. Although dynamic singularities are avoided, the acceleration inputs undergo a sudden amplification when ξ_1 drops close to zero (its minimum positive value is about 0.05 just after $t = 8$ s).

5   Conclusion

In this chapter, two general robotic planning problems have been considered: *(i)* planning a transfer motion between two given configurations for kinematic systems subject to first-order nonholonomic constraints, and *(ii)* planning a rest-to-rest trajectory between two given equilibrium states for dynamic systems subject to second-order nonholonomic constraints.

We have presented planning strategies that rely on two general nonlinear control tools: iterative steering (using nilpotent approximations) and dynamic feedback linearization (or flatness). These solution approaches have been illustrated on non-standard case studies, including two non-flat kinematic systems (the plate-ball manipulation system and the two-trailer mobile robot with non-zero hooking) and two flat dynamic systems (a two-link robot with flexible forearm and a planar underactuated robot with two passive joints).

The proposed methods provide some further benefits from the control point of view. Iterative steering has intrinsic properties of robustness against perturbations. We have shown here that error contraction along the iterations can be enforced also in the presence of uncertainty in the system parameters. The same is clearly true when an exact planner is known for the nominal case (e.g., for a flat or chained-form transformable system), but its iterative application is needed in order to robustify the planner with respect to perturbations (see [40]). Dynamic feedback linearization leads instead to a straightforward (linear) design of a trajectory tracking controller, with global exponential convergence to the planned trajectory when starting with an initial state error (see [17,22]).

From the application point of view, the presented case studies suggest several extensions that need further research. One example is the inclusion of obstacles in a kinematic setting (the complete *motion planning* problem). Noticeably, an advantage of iterative steering is the possibility of shaping the system trajectory during the generic iteration through the choice of an (overparametrized) open-loop command that allows collision avoidance. As for dynamic underactuated robots, the planning problem for systems with degree of underactuation greater than one is still open in general. We have presented a possible two-stage solution for the two-link flexible robot having multiple deformation modes (equal to the degree of underactuation) in its forearm. Indeed, the search for a flat output (if one exists) is a challenging issue in this case, as well as in more general instances of robots with multiple flexible links. Similarly, the removal of the special hinging hypothesis for planar robots with two or more passive joints is of interest. Furthermore, non-planar case studies of underactuated robots are absent in the literature.

Various control theoretical aspects that deserve deeper analysis arise in connection with the presented planning methods for nonholonomically constrained robotic systems: the handling of singularities in the dynamic feedback linearization approach, the use of global non-homogenous nilpotent system approximations, and technical advances in the nilpotent approximation of systems with drift (see [15] for some preliminary results).

References

1. H. Arai, K. Tanie, and N. Shiroma, "Nonholonomic control of a three-dof planar underactuated manipulator," *IEEE Trans. on Robotics and Automation*, vol. 14, pp. 681–695, 1998.
2. E. Barbieri and Ü. Özgüner, "Unconstrained and constrained mode expansions for a flexible slewing link," *ASME J. of Dynamic Systems, Measurement, and Control*, vol. 110, pp. 416–421, 1988.
3. E. Bayo, "A finite-element approach to control the end-point motion of a single-link flexible robot," *J. of Robotic Systems*, vol. 4, pp. 63–75, 1987.
4. A. Bellaïche, "The tangent space in sub-Riemannian geometry," in A. Bellaïche and J.-J. Risler (Eds.), *Sub-Riemannian Geometry*, pp. 1–78, Birkhäuser, 1996.

5. F. Bellezza, L. Lanari, and G. Ulivi, "Exact modeling of the slewing flexible link," *Proc. of 1990 IEEE Int. Conf. on Robotics and Automation*, pp. 734–739, 1990.
6. A. Bicchi and R. Sorrentino, "Dexterous manipulation through rolling," *Proc. of 1995 IEEE Int. Conf. on Robotics and Automation*, pp. 452–457, 1995.
7. W.J. Book, "Modeling, design, and control of flexible manipulator arms: A tutorial review," *Proc. of 29th IEEE Conf. on Decision and Control*, pp. 500–506, 1990.
8. R.W. Brockett and L. Dai, "Non-holonomic kinematics and the role of elliptic functions in constructive controllability," in Z. Li and J. F. Canny (Eds.), *Nonholonomic Motion Planning*, pp. 1–21, Kluwer Academic Publishers, 1993.
9. F. Bullo and K. M. Lynch, "Kinematic controllability for decoupled trajectory planning in underactuated mechanical systems," *IEEE Trans. on Robotics and Automation*, vol. 17, pp. 402–412, 2001.
10. G. Campion, B. d'Andrea-Novel, and G. Bastin, "Modeling and state feedback control of nonholonomic mechanical systems," *Proc. of 30th IEEE Conf. on Decision and Control*, pp. 1184–1189, 1991.
11. A. De Luca and G. Di Giovanni, "Rest-to-rest motion of a one-link flexible arm," *Proc. of 2001 IEEE/ASME Int. Conf. on Advanced Intelligent Mechatronics*, pp. 923–928, 2001.
12. A. De Luca and G. Di Giovanni, "Rest-to-rest motion of a two-link robot with a flexible forearm," *Proc. of 2001 IEEE/ASME Int. Conf. on Advanced Intelligent Mechatronics*, pp. 929–935, 2001.
13. A. De Luca, S. Iannitti, and G. Oriolo, "Stabilization of a PR planar underactuated robot," *Proc. of 2001 IEEE Int. Conf. on Robotics and Automation*, pp. 2090–2095, 2001.
14. A. De Luca and P. Lucibello, "A general algorithm for dynamic feedback linearization of robots with elastic joints," *Proc. of 1998 IEEE Int. Conf on Robotics and Automation*, pp. 504–510, 1998.
15. A. De Luca, R. Mattone, and G. Oriolo, "Stabilization of an underactuated planar 2R manipulator," *Int. J. on Robust and Nonlinear Control*, vol. 10, pp. 181–198, 2000.
16. A. De Luca and G. Oriolo, "Motion planning and trajectory control of an underactuated three-link robot via dynamic feedback linearization," *Proc. of 2000 IEEE Int. Conf. on Robotics and Automation*, pp. 2789–2795, 2000.
17. A. De Luca and G. Oriolo, "Trajectory planning and control for planar robots with passive last joint," *Int. J. of Robotics Research*, vol. 21, pp. 575–590, 2002.
18. A. De Luca, G. Oriolo, and C. Samson, "Feedback Control of a Nonholonomic Car-Like Robot," in J.-P. Laumond (Ed.), *Robot Motion Planning and Control*, pp. 171–253, Springer Verlag, 1998.
19. M. Fliess, J. Lévine, P. Martin, and P. Rouchon, "Flatness and defect of non-linear systems: Introductory theory and examples," *Int. J. of Control*, vol. 61, pp. 1327–1361, 1995.
20. H. Goldstein, *Classical Mechanics*, 2nd Ed., Addison Wesley, 1980.
21. H. Hermes, "Nilpotent and high-order approximations of vector field systems," *SIAM Review*, vol. 33, pp. 238–264, 1991.
22. S. Iannitti and A. De Luca, "Dynamic feedback control of $XYn\overline{R}$ planar robots with n rotational passive joints," *J. of Robotic Systems*, vol. 20, pp. 251–270, 2003.
23. A. Isidori, *Nonlinear Control Systems*, 3rd Ed., Springer Verlag, 1995.
24. V. Jurdjevic, "The geometry of the plate-ball problem," *Arch. for Rational Mechanics and Analysis*, vol. 124, pp. 305–328, 1993.
25. G. Laferriere and H.J. Sussmann, "A differential geometric approach to motion planning," in Z. Li and J. F. Canny (Eds.), *Nonholonomic Motion Planning*, pp. 235–270. Kluwer Academic Publishers, 1992.

26. J.-P. Laumond (Ed.), *Robot Motion Planning and Control*, Springer Verlag, 1998.

27. Z. Li and J. Canny, "Motion of two rigid bodies with rolling constraint," *IEEE Trans. on Robotics and Automation*, vol. 6, pp. 62–72, 1990.

28. D.A. Lizárraga, P. Morin, and C. Samson, *Exponential Stabilization of Certain Configurations of the General N-Trailer System*, Research Report no. 3412, INRIA, 1998.

29. P. Lucibello and G. Oriolo, "Robust stabilization via iterative state steering with an application to chained-form systems," *Automatica*, vol. 37, pp. 71–79, 2001.

30. A. Marigo and A. Bicchi, "Rolling bodies with regular surface: Controllability theory and applications," *IEEE Trans. on Automatic Control*, vol. 45, pp. 1586–1599, 2000.

31. D.J. Montana, "The kinematics of contact and grasp," *Int. J. of Robotics Research*, vol. 7, no. 3, pp. 17–32, 1988.

32. R.M. Murray, "Control of nonholonomic systems using chained forms," *Fields Institute Communications*, vol. 1, pp. 219–245, 1993.

33. R.M. Murray, Z. Li, and S.S. Sastry, *A Mathematical Introduction to Robotic Manipulation*, CRC Press, 1994.

34. R.M. Murray, M. Rathinam, and W. Sluis, "Differential flatness of mechanical control systems: A catalog of prototype systems," *Proc. of 1995 ASME Int. Mechanical Engineering Congr. and Expo.*, 1995.

35. R.M. Murray and S.S. Sastry, "Nonholonomic motion planning: Steering using sinusoids," *IEEE Trans. on Automatic Control*, vol. 38, pp. 700–716, 1993.

36. G. Oriolo, A. De Luca, and M. Vendittelli, "WMR control via dynamic feedback linearization: Design, implementation and experimental validation," *IEEE Trans. on Control Systems Technology*, vol. 10, pp. 835–852, 2002.

37. G. Oriolo and Y. Nakamura, "Control of mechanical systems with second-order nonholonomic constraints: Underactuated manipulators," *Proc. of 30th IEEE Conf. on Decision and Control*, pp. 2398–2403, 1991.

38. G. Oriolo and M. Vendittelli, "Robust stabilization of the plate-ball manipulation system," *Proc. of 2001 IEEE Int. Conf. on Robotics and Automation*, pp. 91–96, 2001.

39. G. Oriolo and M. Vendittelli, *A Stabilization Technique for General Nonholonomic Systems*, DIS Technical Report, Università di Roma "La Sapienza", 2003.

40. G. Oriolo, M. Vendittelli, A. Marigo, and A. Bicchi, "From nominal to robust planning: The plate-ball manipulation system," *Proc. of 2003 IEEE Int. Conf. on Robotics and Automation*, 2003.

41. M. Rathinam and R.M. Murray, "Configuration flatness of Lagrangian systems underactuated by one control," *SIAM J. of Control and Optimization*, vol. 36, pp. 164–179, 1998.

42. P. Rouchon, "Necessary condition and genericity of dynamic feedback linearization," *J. of Mathematical Systems, Estimation and Control*, vol. 4, pp. 257–260, 1994.

43. P. Rouchon, M. Fliess, J. Lévine, and P. Martin, "Flatness, motion planning and trailer systems," *Proc. of 32nd IEEE Conf. on Decision and Control*, pp. 2700–2705, 1993.

44. S. Sekhavat, P. Rouchon, and J. Hermosillo, "Computing the flat outputs of Engel differential systems: The case study of the bi-steerable car," *Proc. of 2001 American Control Conf.*, pp. 3576–3581, 2001.

45. N. Shiroma, H. Arai, and K. Tanie, "Nonholonomic motion planning for coupled planar rigid bodies with passive revolute joints," *Int. J. of Robotics Research*, vol. 21, pp. 563–574, 2002.

46. N. C. Singer and W. P. Seering, "Preshaping command inputs to reduce system vibration," *ASME J. of Dynamic Systems, Measurements, and Control*, vol. 112, pp. 76–82, 1990.

47. M. J. van Nieuwstadt and R. M. Murray, "Real-time trajectory generation for differentially flat systems," *Int. J. of Robust and Nonlinear Control*, vol. 8, pp. 995–1020, 1998.

48. M. Vendittelli and G. Oriolo, "Stabilization of the general two-trailer system," *Proc. of 2000 IEEE Int. Conf. on Robotics and Automation*, pp. 1817–1822, 2000.
49. M. Vendittelli, G. Oriolo, and J.-P. Laumond, "Steering nonholonomic systems via nilpotent approximations: The general two-trailer system," *Proc. of 1999 IEEE Int. Conf. on Robotics and Automation*, pp. 823–829, 1999.

Comparison and Improvement of Control Schemes for Robotic Teleoperation Systems with Time Delay

Paolo Arcara and Claudio Melchiorri

Dipartimento di Elettronica Informatica e Sistemistica
Università di Bologna
Via Risorgimento 2, 40136 Bologna, Italy
<*parcara,cmelchiorri*>@*deis.unibo.it*
http://www-lar.deis.unibo.it

Abstract. Telemanipulation, one of the first area in robotics to be developed, still attracts a noticeable research activity. In particular, several control schemes have been proposed for controlling telemanipulation systems with time delay, a problem that does not seem to be solved yet in a satisfactory manner. In this context, the goals of this chapter are twofold. First, criteria are presented and used in order to analyze and compare control schemes already known in the literature. This study can help the control designer in pointing out the basic features of a given methodology and in the definition of the relative control parameters. Second, a particular control scheme, the so-called IPC (Intrinsically Passive Control), is discussed in detail, and it is shown that proper modifications of the basic scheme can improve its performance in terms of the proposed criteria.

1 Introduction

The capability of operating in remote or hazardous environment is among the main reasons for the development of telemanipulation systems, one of the first area in robotics to be developed [11,15,20,21]. Still today, this field is of relevant interest for the robotics research community, and more and more exciting applications are currently under development: surgery, tele-diagnosis, space operations, security and surveillance, application in mining or wood industry and so on, are just few examples of use of this technology.

Robotic telemanipulation systems have been widely studied in the last decades and, in particular, many different control schemes have been proposed in order to achieve different goals with these devices. One of the main goals is to improve the so-called "transparency" of the system. With this respect, a telemanipulation system should be able to provide the human operator with the capability and the impression of operating "directly" on a remote environment, regardless the presence of the "master" and "slave" devices [12,18,2,22,24].

On the other hand, the level of transparency of a given system is affected by several factors, and the adopted control scheme plays an important role in this scenario. With this respect, one of the goals of this chapter is to present a general framework in which different control schemes for telemanipulation systems can be efficiently compared. One of the results of this comparison is, for example,

B. Siciliano et al. (Eds.): Advances in Control of Articulated and Mobile Robots, STAR 10, pp. 39–60, 2004.
© Springer-Verlag Berlin Heidelberg 2004

an indication on how the control parameters could be tuned for a given scheme. As a case study, six well known control schemes are considered: force reflection, position error, passivity-based force reflection, intrinsically passive controller, and the "four" and "three channels" architectures, see [10,1,8,17,9,23,7,16,14] for a detailed presentation of these controllers.

The comparison of these control schemes is made by considering five criteria, that keep into account both stability and performance, see e.g. [16]. In particular, performance aspects are analyzed in terms of stability, perceived inertia, tracking properties, correct perception of a structured environment, and position drift between the master and slave manipulators. As a consequence of this comparative analysis, the choice of the control parameters in each scheme is briefly discussed, and it is shown that a trade-off between stability and performance naturally arises.

Finally, attention is focused on the intrinsically passive control (IPC) scheme [23,7], and it is shown how a proper modification of the basic control law can lead to significant benefits for some of the proposed indices. In particular, the concept of "variable rest length" of a spring is introduced and used, with a passive approach, for the improvement of the transparency of the whole telemanipulation system.

The structure of this chapter is as follows. Section 2, besides giving the definitions of a typical telemanipulation system with its basic elements, summarizes different control schemes taken from the literature. Section 3 defines the criteria that have been used to compare the control schemes, and shows the comparison results. Section 4 presents a case study, based on the IPC, and illustrates the proposed modifications. Moreover, some simulation results are illustrated and discussed. Finally, Section 5 concludes with some comments and final remarks.

2   Basic Definitions and Control Schemes

This section, after a brief description of a robotic telemanipulation system and the definition of a simple example used as a benchmark in the following comparisons, summarizes the six control schemes, already presented in the literature, that have been considered for the comparative analysis.

In Fig. 1, the main elements and the principal signals of a generic telemanipulation control scheme are schematically illustrated.

For the sake of simplicity, as shown in Fig. 1, it is assumed that the interaction among the different "components" of the telemanipulation scheme is performed by means of the "signals" x (position or velocity, this choice depending on the particular control scheme) and F (force). Note that also the signals in the communication channel depend on the control scheme, as described in the following.

Although in literature methodological tools have been recently proposed for dealing with very complex cases, comprising varying time delays and a full geometrical context, e.g. 6-dimensional architectures ([23,7]), in this chapter a simple example has been used in order to more easily underline the main aspects concerning the choice of a particular control scheme and the selection of its parameters:

- A constant transmission delay T in the communication channel is considered.

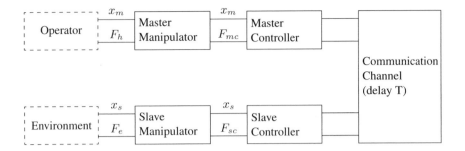

Fig. 1. Basic elements and signals in a teleoperation control scheme; x_i indicates the position or the velocity (depending on the control scheme) of the master/slave system, F_i is a force.

- Both master and slave manipulators are one degree-of-freedom devices, with a linear dynamics (e.g. linearized with proper local controllers) described by:

$$F_m = \left(M_m s^2 + B_m s \right) x_m \tag{1}$$

$$F_s = \left(M_s s^2 + B_s s \right) x_s \tag{2}$$

where M_i, B_i are the manipulator's inertia and damping coefficient respectively, F_i represents the force exerted on the manipulator and x_i its displacement. Subscript $i = m, s$ indicates the master m or slave s manipulator respectively.
- Identical manipulators have been considered at the master and slave side, i.e. $M_m = M_s = M$, and $B_m = B_s = B$.

Although this is a rather simple example, it is sufficient to describe the different characteristics of the considered schemes and help in the definition of the relative control parameters.

The forces applied to both manipulators depend on the "external" interactions (environment and human operator) and on the adopted control scheme. In general, these forces can be written as

$$F_m = F_h - F_{mc} \tag{3}$$

$$F_s = F_e + F_{sc} \tag{4}$$

where F_h, F_e are the forces imposed by the human operator and by the remote environment respectively, while F_{mc}, F_{sc} are the forces computed by the master and slave controllers, see Fig. 2.

In the following, some control schemes taken from the literature are briefly summarized, see [4,5] for further details on these and other control schemes.

2.1 Force Reflection (FR)

Key point of the *force reflection* scheme, one of the first to be proposed for controlling teleoperation systems, is that position information is transmitted from master to slave

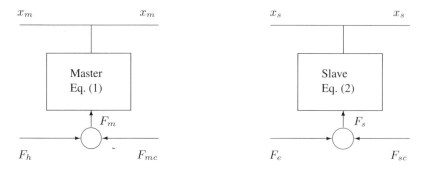

Fig. 2. Forces applied to master and slave (detail of Fig. 1); F_{mc} is usually considered negative.

while force information flows in the opposite direction. The controllers equations are:

$$\begin{cases} F_{mc} = G_m F_{sd} \\ F_{sc} = K_s(x_{md} - x_s) \end{cases} \qquad (5)$$

where K_s, G_m are parameters to be properly tuned. The subscript d indicates the delayed terms connected with the exchanged information between the two sides:

$$\begin{cases} x_{md} = e^{-sT} x_m \\ F_{sd} = e^{-sT} F_{sc} \end{cases} \qquad (6)$$

where e^{-sT} is related to the constant time delay T of the transmission channel.

2.2   Position Error (PE)

The *position error* telemanipulation control scheme is symmetric, since the forces applied to both manipulators depend on the position errors computed on each side. The control equations are:

$$\begin{cases} F_{mc} = K_m(x_m - x_{sd}) \\ F_{sc} = K_s(x_{md} - x_s) \end{cases} \qquad (7)$$

where K_m, K_s are parameters of the control system. In this scheme, position information is exchanged at both sides of the communication channel:

$$\begin{cases} x_{md} = e^{-sT} x_m \\ x_{sd} = e^{-sT} x_s \end{cases} \qquad (8)$$

2.3 Force Reflection with Passivity (FRP)

In order to guarantee *passivity* (and therefore stability) of the *force reflection* scheme, damping injection elements can be added to the control equations, that become:

$$\begin{cases} F_{mc} = G_m F_{sd} + B_b v_m \\ F_{sc} = K_s \left(\dfrac{v_{md} - F_{sc}/B_b}{s} - x_s \right) \end{cases} \tag{9}$$

where v_m represents the velocity of the master manipulator ($v_m = s x_m$) and K_s, G_m, B_b are the parameters of the controllers. The transmitted information is:

$$\begin{cases} v_{md} = e^{-sT} v_m \\ F_{sd} = e^{-sT} F_{sc} \end{cases} \tag{10}$$

2.4 Intrinsically Passive Controller (IPC)

The *intrinsically passive control* scheme is based on passivity concepts and can be interpreted in terms of masses, damping elements and springs, i.e. passive physical components. Therefore, this control scheme can be described as

$$\begin{cases} F_{mc} - F_{mi} = (M_{mc}s^2 + B_{mv}s)x_{mc} \\ F_{mc} = K_{mv}(x_m - x_{mc}) \\ F_{mi} = (K_{mi} + B_{mi}s)(x_{mc} - v_{mi}/s) \\ -F_{sc} - F_{si} = (M_{sc}s^2 + B_{sv}s)x_{sc} \\ F_{sc} = K_{sv}(x_{sc} - x_s) \\ F_{si} = (K_{si} + B_{si}s)(x_{sc} - v_{si}/s) \end{cases} \tag{11}$$

where M_{mc}, B_{mv}, K_{mv}, K_{mi}, B_{mi}, M_{sc}, B_{sv}, K_{sv}, K_{si}, B_{si} are the parameters of the "physical" components; x_{mc}, x_{sc} are the positions of the virtual masses implemented in the controllers; F_{mi}, v_{mi}, F_{si}, v_{si} are the forces and velocities transmitted between master and slave.

In order to guarantee passivity, these signals are combined into *scattering variables*, which are the signals to be exchanged in the communication channel:

$$\begin{cases} S_m^+ = (F_{mi} + B_i v_{mi})/\sqrt{2B_i} \\ S_m^- = (F_{mi} - B_i v_{mi})/\sqrt{2B_i} \\ S_s^+ = -(F_{si} + B_i v_{si})/\sqrt{2B_i} \\ S_s^- = (-F_{si} + B_i v_{si})/\sqrt{2B_i} \end{cases} \tag{12}$$

where B_i represents the impedance of the communication channel. In this case, the signals transmitted between master and slave are

$$\begin{cases} S_s^- = e^{-sT} S_m^+ \\ S_m^- = e^{-sT} S_s^+ \end{cases} \tag{13}$$

2.5 Four Channels (4C)

The name *four channels* indicates that in this particular control scheme four different signals are exchanged in the communication channel. In fact, both velocity and force signals flow between master and slave. The control can be defined as

$$
\begin{cases}
F_{mc} = -C_6 F_h + C_m v_m + F_{sd} + v_{sd} \\
F_{sc} = C_5 F_e - C_s v_s + F_{md} + v_{md}
\end{cases}
\tag{14}
$$

where $v_m = s x_m$, $v_s = s x_s$ represent the velocities, while C_5, C_6 and C_m, C_s are respectively feedforward and feedback dynamic control parameters. The data transmitted in the communication channel are

$$
\begin{cases}
v_{md} = C_1 \, e^{-sT} v_m \\
F_{md} = C_3 \, e^{-sT} F_m \\
F_{sd} = C_2 \, e^{-sT} F_s \\
v_{sd} = C_4 \, e^{-sT} v_s
\end{cases}
\tag{15}
$$

where C_1, C_2, C_3, C_4 are control parameters to be properly selected.

In order to better specify this general control scheme, one can reduce the number of free parameters to two (C_2 and C_3) by setting (see [14]):

$$
\begin{array}{ll}
C_1 = M_s s + B_s + C_s , & C_4 = -(M_m s + B_m + C_m) \\
C_5 = C_3 - 1 , & C_6 = C_2 - 1
\end{array}
\tag{16}
$$

2.6 Three Channels (3C)

A simpler scheme, with only *three channels*, can be obtained from the previous controller by imposing, for example, $C_2 = 0$ in (15) and (16). The control equations are:

$$
\begin{cases}
F_{mc} = F_h + C_m v_m + v_{sd} \\
F_{sc} = (C_3 - 1) F_e - C_s v_s + F_{md} + v_{md}
\end{cases}
\tag{17}
$$

where feedforward actions on F_h, F_e are used. Information flows now in three different channels:

$$
\begin{cases}
v_{md} = C_1 \, e^{-sT} v_m \\
F_{md} = C_3 \, e^{-sT} F_m \\
v_{sd} = C_4 \, e^{-sT} v_s
\end{cases}
\tag{18}
$$

where C_1, C_3, C_4 are parameters that can be set as specified in (16).

3   Comparison Criteria and Results

It is of interest to establish general criteria by means of which control schemes for telemanipulation systems can be evaluated and compared. These criteria should consider the performance achieved by the different schemes. In particular, five different aspects are considered here:

1. stability,
2. inertia and damping,
3. tracking,
4. stiffness,
5. drift.

Further details concerning the comparison of these and other telemanipulation control schemes, the tuning of the parameters and the maximum performance obtainable with each scheme can be found in [5,3]. As shown in these works, stability and performance are always conflictual aspects, and the choice of the control parameters is often the result of a trade-off between them.

CR1. Stability The stability of a telemanipulation scheme is strongly related to the amount of time delay T in the transmission channel. In practice, one can enumerate two main cases:

1. **IS** schemes which are *intrinsically stable* (IS), that is stability is automatically guaranteed independently of time delay T.
2. **PS** schemes which are *possibly stable* (PS), i.e. that can be rendered stable, for any value of the delay T, with a proper choice of the controller's parameters.

Table 1 shows the stability properties of the control schemes summarized in the previous section. It is worth noticing that four- and three-channels control schemes are usually of PS type but, due to the choice of the parameters in Eq. (16), intrinsic stability independently of time delay T can be achieved. More details on this topic can be found in [5].

Table 1. Stability properties of the considered schemes.

	FR	PE	FRP	IPC	4C	3C
IS			•	•		
PS	•	•			•	•

CR2. Inertia and Damping These aspects are related to the perception of the user while moving the master manipulator when the remote arm is not in contact with the environment. In this case, the inertia and damping perceived at the master can be described by means of the following transfer function:

$$G_1(s) \equiv \left(\frac{x_m}{F_h} \bigg|_{F_e=0} \right)^{-1} \tag{19}$$

In order to compute the inertia and damping, one can rewrite (19) as

$$G_1(s) = M_{eq}s^2 + B_{eq}s + G_1^*(s) \tag{20}$$

where M_{eq}, B_{eq} represent the parameters under consideration (inertia and damping), and $G_1^*(s)$ contains negligible terms of third and higher order, satisfying the condition $\lim_{s \to 0} G_1^*(s)/s^2 = 0$.

Table 2 contains the expressions of the perceived inertia M_{eq} and damping B_{eq} for the considered control schemes.

Table 2. Inertia and damping terms for the considered control schemes.

Scheme	Inertia (M_{eq})	Damping (B_{eq})
FR	$(1+G_c)M_m - G_cB_m\left(\frac{B_m}{K_c} + 2T\right)$	$(1+G_c)B_m$
PE	$2(M_m - B_mT - K_cT^2) - \frac{B_m^2}{K_c}$	$2(B_m + K_cT)$
FRP	$\left(1 + \frac{G_cB_i^2}{(B_m+B_i)^2}\right)M_m - \frac{2G_cB_iB_mT}{B_m+B_i} - \frac{G_cB_i^2B_m^2}{(B_m+B_i)^2K_c}$	$B_m + B_i + \frac{G_cB_iB_m}{B_m+B_i}$
IPC	$2(M_m + M_{mc}) + B_iT - \frac{(B_m + B_{mc})^2T}{B_i} + \dots$ $\dots - \frac{2B_{mc}^2(2K_{mi} + K_{mc})}{K_{mi}K_{mc}} + \dots$ $\dots - \frac{2B_m(K_{mi} + K_{mc})(B_m + 2B_{mc})}{K_{mi}K_{mc}}$	$2(B_m + B_{mc})$
4C	$\frac{2T(B_m + C_m)}{C_2 + C_3}$	0
3C	$\frac{2T(B_m + C_m)}{C_3}$	0

CR3. Tracking An important property of a telemanipulation system is the ability of the slave device, when it is not in contact with the environment, to track as closely as possible the movements of the master. The tracking properties can be expressed by the following transfer function

$$G_2(s) \equiv \left. \frac{x_m - x_s}{F_h} \right|_{F_e=0} \tag{21}$$

Also in this case it is convenient to identify a constant term δ, that represents the steady-state error between master and slave positions as a consequence of a unit step of the input force F_h:

$$G_2(s) = \delta\, G_2^*(s) \tag{22}$$

where $G_2^*(s)$ satisfies $G_2^*(0) = 1$ and therefore, in steady state conditions, one obtains $G_2(0) = \delta$.

Table 3 reports the values of the tracking error δ for the different schemes. It is important to note that in the FRP case one obtains the velocity error $\delta_v = \frac{B_m}{B_b^2 + B_m^2 + B_b B_m (2 + G_m)}$. As a consequence, the tracking position error is limited only if $B_m = 0$, in which case $\delta = \frac{M_m + B_b T}{B_b^2}$.

Table 3. Tracking errors for the different control schemes.

Scheme	Tracking (δ)
FR	$\dfrac{B_m + K_c T}{B_m K_c (1 + G_c)}$
PE	$\dfrac{1}{2 G_c K_c}$
FRP	∞
IPC	$\dfrac{2 B_i (K_{mi} + K_{mc}) + K_{mi} K_{mc} T}{2 B_i K_{mi} K_{mc}}$
4C	$\dfrac{C_2 - C_3}{2(B_m + C_m)}$
3C	$\dfrac{-C_3}{2(B_m + C_m)}$

CR4. Stiffness Another important aspect for the evaluation of the performance of a telemanipulation scheme is the correct perception, for the human operator, of the stiffness of the remote environment. Assuming for example the case of interaction with an environment with known stiffness K_e and damping B_e, the perceived stiffness can be measured with

$$G_3(s) \equiv \left(\left. \frac{x_m}{F_h} \right|_{F_e=-(B_e s+K_e)x_s} \right)^{-1} \tag{23}$$

In this case, one can identify a constant term K_{eq} that represents the perceived stiffness:

$$G_3(s) = K_{eq}G_3^*(s) \tag{24}$$

where $G_3^*(s)$ satisfies $G_3^*(0) = 1$. Table 4 reports, in the first column, the resulting stiffness values for the different schemes. It is worth noticing that, for the FRP scheme, one perceives no stiffness ($K_{eq} = 0$) and only a damping factor equal to $B_{eq} = B_b + B_m + B_b G_m$.

Table 4. Perceived stiffness for the control schemes.

Scheme	Stiffness (K_{eq})
FR	$\dfrac{K_e G_c K_c}{K_e + K_c}$
PE	$\dfrac{K_e G_c K_c}{K_e + K_c}$
FRP	0
IPC	$\dfrac{K_e B_i K_{mi} K_{mc}}{B_i(K_{mi}K_{mc} + 2K_e(K_{mi} + K_{mc})) + K_e K_{mi} K_{mc} T}$
4C	$\dfrac{K_e(C_2 + C_3)(B_m + C_m)}{(C_2 + C_3)(B_m + C_m) + 2K_e C_2 C_3 T}$
3C	K_e

CR5. Drift The last parameter to be evaluated is the position drift between the manipulators. This parameter is similar to the tracking error, the only difference being the interaction, at the slave side, with a structured environment with stiffness K_e and damping B_e. The following transfer function is used to evaluate the position drift:

$$G_4(s) \equiv \left. \frac{x_m - x_s}{F_h} \right|_{F_e=-(B_e s+K_e)x_s} \tag{25}$$

Again, one can identify a constant term that represents the position drift between the master and slave displacements:

$$G_4(s) = \Delta \, G_4^*(s) \tag{26}$$

where $G_4^*(s)$ satisfies $G_4^*(0) = 1$. Table 5 reports, in the second column, the values of the position drift Δ for the considered control schemes. In the FRP case, one obtains a velocity drift $\Delta_v = \dfrac{1}{B_b + B_m + B_b G_m}$, that in general generates an unlimited position drift.

Table 5. Perceived drift for the control schemes.

Scheme	Drift (Δ)
FR	$\dfrac{1}{G_c K_c}$
PE	$\dfrac{1}{G_c K_c}$
FRP	∞
IPC	$\dfrac{2B_i(K_{mi} + K_{mc}) + K_{mi}K_{mc}T}{B_i K_{mi}K_{mc}}$
4C	$\dfrac{2C_2 C_3 T}{(C_2 + C_3)(B_m + C_m)}$
3C	0

4 The IPC

In this section, attention is focused on the Intrinsically Passive Controller (IPC) discussed in 2.4. Certainly, this is not the first control scheme inspired to passivity criteria to be proposed for telemanipulation, see e.g. [18] or the more recent contributions presented in [13,19]. On the other hand, being inspired on physical intuition and analogy, the IPC can be used as a general framework for developing passive controllers and, moreover, can be easily extended to the full (3D) context. This latter aspect, besides representing an elegant mathematical feature, represents a powerful property able to guarantee in non-trivial contexts (as in case e.g. of different dimensions/kinematics of the master-slave devices) the stability of the overall system. This can be achieved also with other schemes, although in a less elegant and more complicated manner. Goal of this section is to present a proper modification of the basic IPC algorithm which improves the performance, in terms of the above criteria, of the whole telemanipulation system.

As already mentioned, the IPC scheme can be described by means of physical elements. Figure 3 describes the master controller as composed by a virtual mass connected, through springs and dampers, to the master manipulator's end-effector on one side (Port 1) and to a power port to the remote slave controller on the opposite side (Port 2). Port 2, defined on both master and slave sides, exchanges power (i.e. forces and velocities) via the scattering (or wave) variables transmitted in the communication channel, characterized by the time delay T.

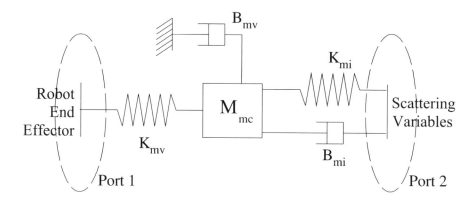

Fig. 3. Physical description of the Intrinsically Passive Controller (master side).

In Section 3, *tracking* and *drift* have been presented as important criteria for evaluating telemanipulation performance. In the following, these indices are used in a slightly more general context. In fact, by considering an IPC scheme active at both master and slave side, the achievable performance can be evaluated by considering the difference $\Delta(t) = x_m(t) - x_s(t)$ between the master and slave manipulator displacements in three different working situations:

A. No interaction with the environment ($F_e = 0$, i.e. no force is applied to the slave manipulator), this case is related to the analysis of the *tracking* properties, see *CR3*.
B. Interaction with a stiff environment (F_e proportional to the displacement of the slave manipulator), i.e. related to the analysis of the *drift* criterion, see *CR5*.
C. Interaction with a viscous environment (F_e proportional to the velocity of the slave manipulator).

In order to evaluate the performance of the IPC scheme in these cases, simulations have been carried out considering a constant external force applied by the human operator, i.e. $F_h(t) =$
$$\bar{F}_h = 10.$$

The following values for the manipulators and IPC parameters have been chosen for this case study (dimensions are expressed in the standard units: Kg, m, s):

$$M = 10, \ B = 1,$$
$$M_{mc} = M_{sc} = 1,$$
$$B_{mv} = B_{sv} = 20,$$
$$K_{mv} = K_{sv} = 500,$$
$$K_{mi} = K_{si} = 500,$$
$$B_{mi} = B_{si} = 20, \ B_i = 20.$$

The time delay has been fixed to $T = 0.2 \ s$.

The simulation results, summarized in Fig. 4, are commented in the following.

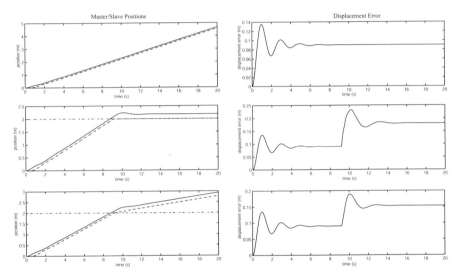

Fig. 4. Master (solid) and slave (dashed) displacements [left] and displacement error [right] in the case of no environment interaction [top], stiff environment [center] and viscous environment [bottom]; the position of the environment is $x_e = 2 \ m$.

Case A: No Environment If no external force is applied to the the slave manipulator (i.e. $F_e(t) = 0$), one can compute the steady-state tracking error, see CR3, as:

$$\bar{\Delta}_A = \frac{2B_i(K_{mi} + K_{mv}) + TK_{mi}K_{mv}}{2B_iK_{mi}K_{mv}} \bar{F}_h = \delta \bar{F}_h \tag{27}$$

The constant term

$$\delta = \frac{2B_i(K_{mi} + K_{mv}) + TK_{mi}K_{mv}}{2B_iK_{mi}K_{mv}} \tag{28}$$

depends on the parameters B_i, K_{mi} and K_{mv}, and on the time delay T. Figure 4 [top] shows the master and slave displacements in this case. One can observe that, as a consequence of the constant force $\bar{F}_h$, a constant velocity on both sides is obtained because of the presence of the dissipative elements (dampers).

Case B: Stiff Environment The interaction with a stiff environment (modeled as $F_e(t) = -K_e[x_s(t) - \bar{x}_e]$, where $\bar{x}_e$ is the position of the environment) at the remote side produces a zero steady-state velocity, with the following drift:

$$\bar{\Delta}_B = \frac{2B_i(K_{mi} + K_{mv}) + TK_{mi}K_{mv}}{B_iK_{mi}K_{mv}}\bar{F}_h = 2\delta\bar{F}_h = \Delta\bar{F}_h \tag{29}$$

see CR5. Note that $\bar{\Delta}_B$ does not depend on the stiffness coefficient K_e of the environment. Figure 4 [center] shows master and slave positions when $K_e = 1 \cdot 10^6$ and $\bar{x}_e = 2$. One can observe that, after a transient period, a zero velocity condition is obtained on both sides.

Case C: Viscous Environment In case of viscous environment (i.e. $F_e(t) = -B_e\dot{x}_s(t)$) at the slave side, a constant steady-state velocity is obtained. In this case, the difference between the manipulators displacements is:

$$\bar{\Delta}_C = \frac{(B + B_{mv} + B_e)(2B_i(K_{mi} + K_{mv}) + TK_{mi}K_{mv})}{(2B + 2B_{mv} + B_e)B_iK_{mi}K_{mv}}\bar{F}_h \tag{30}$$

and, using again the constant term δ introduced in CR3 and defined in (28), one obtains the following simplified expression:

$$\bar{\Delta}_C = \frac{2(B + B_{mv} + B_e)}{2(B + B_{mv}) + B_e}\delta\bar{F}_h \tag{31}$$

Figure 4 [bottom] shows master and slave manipulators displacements when $B_e = 100$. Finally, one can easily verify from (27) and (29) that Case A and Case B, i.e. without interaction with the environment and with a stiff environment, are particular instances of Case C with, respectively, $B_e = 0$ and $B_e = +\infty$.

4.1   Variable Rest Length Springs

As shown previously, an IPC scheme introduces a displacement error between master and slave. A possible solution in order to reduce, or even completely avoid, this drawback is to introduce a proper modification, through additional passive terms, in both the master and slave controllers. For this purpose, a novel type of "spring" is here introduced for the IPC scheme. Main feature of this new spring is the possibility of changing the "rest length", i.e. the configuration for which it does not generate any force. In summary, the variations in the rest lengths of the springs are used as an efficient method to reduce position errors and achieve good transparency properties. This method guarantees also passivity, as discussed in the following, and can easily be implemented through the addition of an energetic port for each spring.

As well known, the relationship between force and displacement of a linear spring is $F = K(x_1 - x_2)$, where the force F depends on both the stiffness coefficient

K and the difference between the positions x_1, x_2 of the points connected by the spring.

In order to consider a varying rest length for the spring, one can consider $F = K(x_1 - x_2 + x_l)$, where the new term x_l represents the rest length of the spring. In practice, an additional energetic port is defined for the springs, being this port described as

$$\begin{cases} F_l = K(x_1 - x_2 + x_l) \\ v_l = \dot{x}_l \end{cases}$$

where F_l, v_l are the force and velocity associated to the new energetic port.

The potential energy of the spring and the overall power exchange can be written as

$$\begin{cases} E_p = \frac{K}{2}(x_1 - x_2 + x_l)^2 \\ P_{tot} = F_1 v_1 - F_2 v_2 + F_l v_l = \frac{d}{dt}E_p \end{cases}$$

These equations describe a three-port system. The first two ports are related to the power exchanged with the two points connected via the spring, as in the normal case, while the last port is associated to the variations of the rest length; v_1, v_2 are the velocities of the linked objects (i.e. of the two ends of the spring), and $F_1 = F_2 = F$ are the forces exerted on the two objects. The equations state that the system is lossless or, equivalently, that is passive without any power dissipation.

Figure 5 shows two springs with positive and negative rest length value and the bond graph of a simple variable rest length spring.

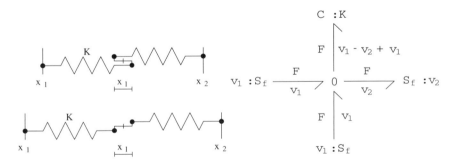

Fig. 5. Springs with variable rest length: case $x_l > 0$ (upper part) and case $x_l < 0$ (lower part) [left] and bond graph of a spring with variable rest length [right].

4.2   Proposed Control Algorithm

A modification is now introduced in the IPC scheme with the aim of improving the performance of the whole system. The proposed solution compensates for the difference between the displacements of the two manipulators by varying the rest lengths of the master and slave controller springs. For this purpose, a *supervisor* is inserted for each variable spring in order to compute the correct rest lengths to be imposed. In this manner, one obtains an adaptive controller, whose parameters dynamically change according to the different environments one has to interact with.

Main goal of both the master and slave supervisors is to measure or compute the force applied by the human operator (F_h) and the characteristic impedance of the environment (B_e, varying from zero to infinity), in order to apply the necessary compensation to counteract the displacement error.

The supervisors receive as inputs the position and force measurements. Moreover, the time delay T has to be known. In fact, as shown in (30), the difference between the manipulator displacements depends on these factors as well as on the parameters of the controllers.

The two supervisors have to compute and, possibly, compensate for the displacement difference $\Delta(t)$, defined in (31), on each of the two sides of the telemanipulation scheme. Therefore, the error term to be compensated on each side is half the total error:

$$\tilde{\Delta}(t) = \frac{1}{2}\bar{\Delta}_C = \frac{B + B_{mv} + \tilde{B}_e(t)}{2(B + B_{mv}) + \tilde{B}_e(t)} \, \delta \, \tilde{F}_h(t) \tag{32}$$

where $\tilde{x}$ is the estimation of the (unknown) variable x.

The two supervisors act according to the following main steps:

1. Identification of the force $\tilde{F}_h(t)$ applied by the operator.
2. Identification of the characteristic impedance $\tilde{B}_e(t)$ of the environment.
3. Computation of half of the displacement error $\tilde{\Delta}(t)$ according to Eq. (32).
4. Local compensation at both the master and slave side of $\tilde{\Delta}(t)$ by properly setting the rest length of the spring.

These steps must be carried out differently on each side of the scheme. Since the telemanipulation system under study is symmetric, one can define as master side the part of the scheme in which power is inserted from the human operator and, on the converse, slave side is the side in which power is extracted from the external environment. These two distinct cases can be simply discriminated by observing the sign of the products $F_h(t)\dot{x}_m(t)$ and $F_e(t)\dot{x}_s(t)$, representing the power exchanged on the two sides.

Master Supervisor The master supervisor has to estimate half of the displacement error to be compensated, and needs the following variables: the force applied by the human operator (F_h), the local manipulator velocity ($\dot{x}_m$), the transmission time delay (T). Moreover, it needs also the control parameters. The above steps become now:

1) Identification of the applied external force: not necessary in this case because the information is directly available ($\tilde{F}_h(t) = F_h(t)$).

2) The manipulator velocity $\dot{x}_m(t)$ is compared with a threshold velocity $v^* = \varepsilon(> 0)$:

case 2.a)   if $|\dot{x}_m(t)| \leq v^*$
            then $\tilde{B}_e(t) = +\infty$

case 2.b)   if $|\dot{x}_m(t)| > v^*$
            then $\tilde{B}_e(t) = \max\{F_h(t)/\dot{x}_m(t) - 2B - 2B_{mv}, 0\}$

In fact, the impedance perceived at the master side is $B_e + 2B + 2B_{mv}$ and must be corrected because of the presence of the two manipulators and IPCs in the telemanipulation scheme.

3) Computation of half of the displacement error leads to:

case 2.a) $\tilde{\Delta}_m(t) = \delta\, F_h(t)$ (33)

case 2.b) $\tilde{\Delta}_m(t) = \delta\, \max\{F_h(t) - (B + B_{mv})\dot{x}_m(t), F_h(t)/2\}$ (34)

These values are obtained from (32) by substitution of the values of $\tilde{F}_h(t)$ and $\tilde{B}_e(t)$.

4) The rest length of the spring in the master controller is set to $x_{ml}(t) = \tilde{\Delta}_m(t)$, and therefore the "force" computed and applied to the master by the IPC is

$$F_{mc}(t) = K_{mv}(x_m(t) - x_{mc}(t) + \tilde{\Delta}_m(t))$$ (35)

Slave Supervisor The slave supervisor acts on different variables with respect to the master supervisor. The variables of interest at the slave side, besides the controller parameters, are: the environment force F_e, the local manipulator velocity $\dot{x}_s$, the communication time delay T. The above steps become now:

1) Estimation of the external force

$$\tilde{F}_h(t) = F_e(t) - 2B\dot{x}_s(t) - 2B_{mv}\dot{x}_s(t)$$

in fact, the force perceived at the slave is $F_h - 2Bv - 2B_{mv}v$, where v represents the velocity, because part of the force applied by the operator is compensated by the damping terms in the two IPCs.

2) The velocity $\dot{x}_s(t)$ is compared with a threshold velocity $v^* = \varepsilon\ (> 0)$:

case 2.a)   if $|\dot{x}_s(t)| \leq v^*$
            then $\tilde{B}_e(t) = +\infty$

case 2.b)   if $|\dot{x}_s(t)| > v^*$
            then $\tilde{B}_e(t) = -F_e(t)/\dot{x}_s(t)$

3) Computation of half of the displacement error:

case 2.a $\tilde{\Delta}_s(t) = \delta F_e(t)$ (36)

case 2.b $\tilde{\Delta}_s(t) = \delta(F_e(t) - (B + B_{mv})\dot{x}_s(t))$ (37)

where, as for the master supervisor, these equations are obtained from (32).

4) The rest length of the spring of the slave controller is set to $x_{sl}(t) = \tilde{\Delta}_s(t)$ and therefore

$$F_{sc}(t) = K_{sv}(x_s(t) - x_{sc}(t) + \tilde{\Delta}_s(t)) \tag{38}$$

Finally, it is worth noticing that:

1. The exact knowledge of *parameter B* of both manipulators is not necessary in the computations (since the same dynamics is considered at both sides). In fact, by setting $B = 0$ in the master and slave supervisors equations, manipulators viscosity is considered as part of the estimated environment impedance $\tilde{B}_e(t)$, thus leading indirectly to a correct compensation of the position error.
2. Compensation of the displacement error via springs' rest lengths has been expressed as a *linear function* of force and velocity terms with properly defined coefficients, i.e. $\tilde{\Delta}_m = C_{m1}F_h + C_{m2}\dot{x}_m$ and $\tilde{\Delta}_s = C_{s1}F_e + C_{s2}\dot{x}_s$, for apposite values of the coefficients C_{m1}, C_{m2}, C_{s1} and C_{s2}.

4.3 Passivity Considerations

The proposed solution can destabilize the whole telemanipulation system if passivity is not maintained. Note that the proposed adaptive control, based on the variation of the springs rest lengths, does not maintain *a priori* the passivity property. As a matter of fact, since part of the telemanipulation system can be modeled as shown in Fig. 6 [left], the supervisor could inject power into the system thus leading to a non-passive system.

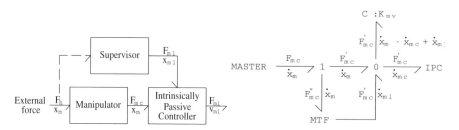

Fig. 6. Master side of the telemanipulation scheme with supervisor control [left] and bond graph representation of the passive supervisor control [right].

The possible lost of passivity can also be pointed out by representing in state space the master side of the telemanipulation scheme:

$$\begin{bmatrix} \dot{x}_m \\ F_{mc} \end{bmatrix} = \begin{bmatrix} 0 & 1 & 0 \\ K_{mv} & K_{mv}C_{m2} & -K_{mv} \end{bmatrix} \begin{bmatrix} x_m \\ \dot{x}_m \\ x_{mc} \end{bmatrix} + \begin{bmatrix} 0 & 0 \\ K_{mv}C_{m1} & 0 \end{bmatrix} \begin{bmatrix} F_h \\ -\dot{x}_{mc} \end{bmatrix}$$

The power injected in this subsystem can be expressed by the inner product of $[\dot{x}_m^T \ F_{mc}^T]^T$ and $[F_h^T - \dot{x}_{mc}^T]^T$. By considering the matrix $D = \begin{bmatrix} 0 & 0 \\ K_{mv}C_{m1} & 0 \end{bmatrix}$, expressing the direct connection between them, one can easily see that $D + D^T$ is not positive definite, and therefore conclude that the system is not passive.

On the other hand, passivity can be obtained by taking directly into account passivity requirements. Main idea is that energy storage in the spring rest length can be allowed only if external power is injected either from the human operator or from the environment, otherwise one could create energy into the system thus leading to a non passive scheme. This can be described via a bond graph, as shown in Fig. 6 [right].

From a mathematical point of view, this idea can be described as:

$$\begin{cases} F_{mc}(t) = K_{mv}(x_m(t) - x_{mc}(t) + x_{ml}(t))(1 + \dot{x}_{ml}(t)/\dot{x}_m(t)) \\ F_{sc}(t) = K_{sv}(x_s(t) - x_{sc}(t) + x_{sl}(t))(1 + \dot{x}_{sl}(t)/\dot{x}_s(t)) \end{cases} \quad (39)$$

where the new variables $\dot{x}_{il}(t)$ with $i = m, s$ (variation of the springs rest length) have to be properly defined in order to achieve the best compensation of the displacement error that is compatible with passivity constrains. In fact, ideally one should have $x_{il} = \tilde{\Delta}_i$ at each time instant; in practice, one can use a proportional controller to define the value $\dot{x}_{il} = K_p(\tilde{\Delta}_i - x_{il})$.

In order to attenuate the effects of the terms $\dot{x}_{il}/\dot{x}_i, i = m, s$ appearing in the new master and slave controllers, see (39), one can define $\dot{x}_{il}(t)$ as a suitable function of $\dot{x}_i(t)$ in order to cancel the terms $\dot{x}_i$, e.g.

$$\begin{cases} \dot{x}_{ml}(t) = K_p(\tilde{\Delta}_m(t) - x_{ml}(t))\dot{x}_m^2(t) \\ \dot{x}_{sl}(t) = K_p(\tilde{\Delta}_s(t) - x_{sl}(t))\dot{x}_s^2(t) \end{cases} \quad (40)$$

where the terms $\dot{x}_i^2(t), i = m, s$, have been used to obtain $F_{ic}(t) = F'_{ic}(t)$ when the manipulator's velocities are zero.

4.4 Simulation Results

In order to evaluate the presented solution, three kinds of simulations, related to the situations described in the case study above, have been carried out with the new controllers.

The results are shown in Fig. 7 for the first controller, Eq. (33)–(38).

These simulations confirm the theoretical results of perfect compensation of the displacement error between master and slave displacements by using varying rest length springs in the IPCs. In general, transient periods depend on the parameters of the IPC and can be further improved. Note that this control scheme does not preserve the overall passivity.

In the second approach, in order to maintain passivity the supervisor energetic port can inject or extract only a finite amount of energy according to the different working situations. The performance are worse than the previous ones (the value $K_p = 500$ has been used), but passivity is now preserved. Note that the displacement error is not completely canceled and that transient periods are longer than in the previous case. However, performance are improved with respect to the original IPC telemanipulation scheme. Figure 8 shows the results for the passive controller described in (39), (40).

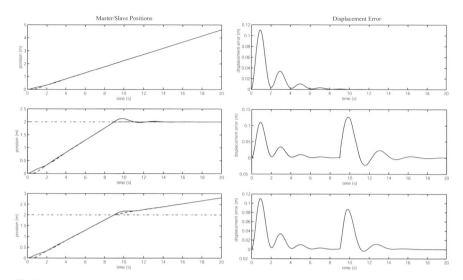

Fig. 7. Master (solid) and slave (dashed) displacements [left] and displacement error [right] in case of no interaction [top], stiff environment [center] and viscous environment [bottom].

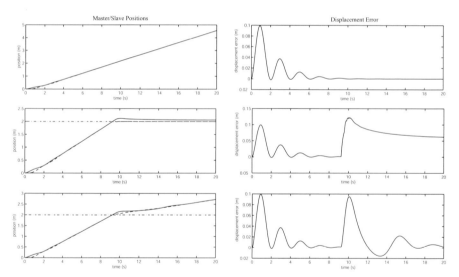

Fig. 8. Master (solid) and slave (dashed) displacements [left] and displacement error [right] with the passive approach in the case of no interaction [top], stiff environment [center] and viscous environment [bottom].

5   Conclusion

In this chapter, a comparison study of several telemanipulation control schemes taken from the literature and based on different control concepts has been presented. The study has been carried out by considering different aspects: both stability and performance have been analyzed, with particular attention to the perceived inertia and damping, tracking, perceived stiffness and position drift. To this regard, comparison criteria have been defined and introduced.

Besides the comparison of control schemes, this study has been proved useful also for the improvement of their performance. In particular, the IPC has been considered, showing that this control scheme can be improved by properly using variable rest length springs. These components are used in such a way that the passivity of the overall scheme is guaranteed and the proposed criteria optimized.

Future work will address the fact that, for the basic IPC scheme, knowledge of force measures and of the (constant) time delay T (as assumed here) is not necessary. Therefore, also for the proposed controller this property should hold. Moreover, nonlinear teleoperation systems will be considered, as well as further comparison criteria and methods to enhance performance.

Acknowledgement

This work has been co-funded by contract IST-2001-34166 (GeoPlex).

References

1. R.J. Anderson and M.W. Spong, "Bilateral control of teleoperators with time delay," *IEEE Trans. on Automatic Control*, vol. 34, pp. 494–501, 1989.
2. R.J. Anderson and M.W. Spong, "Asymptotic stability for force reflecting teleoperators with time delay," *Int. J. of Robotics Research*, vol. 11, pp. 135–149, 1992.
3. P. Arcara, *Control of Haptic and Robotic Telemanipulation Systems*, Ph.D. Thesis, DEIS, Univ. of Bologna, 2002.
4. P. Arcara and C. Melchiorri, "A comparison of control schemes for teleoperation with time delay," *Proc. of IFAC Conf. on Telematics Applications in Automation and Robotics*, pp. 505–510, 2001.
5. P. Arcara and C. Melchiorri, "Control schemes for teleoperation with time delay: A comparative study," *Robotics and Autonomous Systems*, vol. 38, pp. 49–64, 2002.
6. P. Arcara and C. Melchiorri, "Position drift compensation for a passivity-based telemanipulation control scheme," *Proc. of Mechatronics Forum*, pp. 933–943, 2002.
7. P. Arcara, C. Melchiorri and S. Stramigioli, "Intrinsically passive control in bilateral teleoperation MIMO systems," *Proc. of 6th European Control Conf.*, pp. 1180–1185, 2001.
8. A. Eusebi, *Sistemi di Teleoperazione Bilaterale di Posizione e Forza in Presenza di Ritardi*, Ph.D. Thesis, DEIS, Univ. of Bologna, 1995.
9. A. Eusebi and C. Melchiorri, "Force reflecting telemanipulators with time-delay: Stability analysis and control design," *IEEE Trans. on Robotics and Automation*, vol. 14, pp. 635–640, 1998.

10. W.R. Ferrell, "Remote manipulation with transmission delay," *IEEE Trans. on Human Factors Electronics*, vol. 8, pp. 449–455, 1996.

11. R.C. Goertz and M.W. Thompson, "Electronically controlled manipulators," *Nucleonics*, vol. 12, pp. 46–47, 1954.

12. B. Hannaford, "Stability and performance tradeoffs in bi-lateral telemanipulation," *Proc. of 1989 IEEE Int. Conf. on Robotics and Automation*, pp. 1764–1767, 1989.

13. B. Hannaford and J.-H. Ryu, "Time-domain passivity control of haptic interfaces," *IEEE Trans. on Robotics and Automation*, vol. 18, pp. 1–18, 2002.

14. K. Hashtrudi-Zaad and S.E. Salcudean, "On the use of local force feedback for transparent teleoperation," *Proc. of 1999 IEEE Int. Conf. on Robotics and Automation*, pp. 1863–1869, 1999.

15. A. Kobrinskii, "The thought control the machine: Development of a bioelectric prosthesis," *Proc. of 1st IFAC World Congress on Automatic Control*, 1960.

16. D.A. Lawrence, "Stability and transparency in bilateral teleoperation," *IEEE Trans. on Robotics and Automation*, vol. 9, pp. 624–637, 1993.

17. C. Melchiorri and A. Eusebi, "Telemanipulation: System aspects and control issues," in C. Melchiorri and A. Tornambè (Eds.), *Modelling and Control of Mechanisms and Robots*, pp. 149–183, World Scientific Publishers, 1996.

18. G. Niemeyer and J.E. Slotine, "Stable adaptive teleoperation," *Int. J. of Oceanic Engineering*, vol. 16, pp. 152–162, 1991.

19. J.-H. Ryu, D.-S. Kwon and B. Hannaford, "Stable teleoperation with time-domain passivity control," *Proc. of 2002 IEEE Int. Conf. on Robotics and Automation*, pp. 3260–3265, 2002.

20. T.B. Sheridan, "Telerobotics," *Automatica*, vol. 25, pp. 487–507, 1989.

21. T.B. Sheridan, *Telerobotics, Automation, and Human Supervisory Control*, MIT Press, 1992.

22. T.B. Sheridan, "Space teleoperation through time delay: Review and prognosis," *IEEE Trans. on Robotics and Automation*, vol. 9, pp. 592–606, 1993.

23. A.J. van der Schaft, B.M. Maschke, S. Stramigioli, S. Andreotti, and C. Melchiorri, "Geometric scattering in telemanipulation of generalized port controlled Hamiltonian systems," *Proc. of 39th IEEE Conf. on Decision and Control*, pp. 5108–5113, 2000.

24. Y. Yokokohji and T. Yoshikawa, "Bilateral control of master-slave manipulators for ideal kinesthetic coupling — Formulation and experiment," *IEEE Trans. on Robotics and Automation*, vol. 10, pp. 605–620, 1994.

Measuring and Improving Performance in Anti-Windup Laws for Robot Manipulators

Federico Morabito[1], Salvatore Nicosia[1], Andrew R. Teel[2], and Luca Zaccarian[1]

[1] Dipartimento di Informatica Sistemi e Produzione
   Università di Roma Tor Vergata
   Via del Politecnico 110, 00133 Roma, Italy
   <nicosia,zack>@disp.uniroma2.it
   http://www.disp.uniroma2.it
[2] Department of Electrical and Computer Engineering
   University of California
   Santa Barbara, CA 93106, USA
   teel@ece.ucsb.edu
   http://www.ece.ucsb.edu/ccec

Abstract. In this chapter we provide a high performance solution to the anti-windup problem for control systems of robot manipulators undergoing actuator torque saturation. Based on the preliminary work of [10], we provide here improved anti-windup laws based on simple and intuitive parameter tuning. Global asymptotic (and local exponential) stability of the arising closed loops is formally proven for set-point regulation tasks and demonstrated on a simulation example. The simulation examples also show dramatic improvements as compared to previous results.

1   Introduction

Actuator saturation is one of the most common unmodeled phenomena in classical control systems. One of the most studied fields where actuator saturation is involved is that of linear control systems for linear plants. In particular, in the past years a great deal of attention has been given to the study of the so-called "windup" problem for linear plants, wherein a predesigned linear controller is known to work very desirably when interconnected to the linear plant but unpredictable behavior and, often, instability occurs if the input saturation effect is taken into account when interconnecting the controller to the plant. For these windup-prone control systems, "anti-windup design" denotes the synthesis of suitable (linear or nonlinear) filters which augment the original linear controller with the goal of:

1. preserving the linear response prespecified by the linear closed loop as long as the saturation limits are never reached by the actuators;
2. guaranteeing as much as possible the recovery of this linear closed-loop response for all other trajectories.

Many useful constructions are nowadays available in the literature for linear anti-windup designs (see, e.g., [4,8,3] for some recent surveys).

B. Siciliano et al. (Eds.): Advances in Control of Articulated and Mobile Robots, STAR 10, pp. 61–85, 2004.
© Springer-Verlag Berlin Heidelberg 2004

A parallel reasoning can be made when dealing with more complicated control systems, such as a nonlinear controller interconnected to a robotic manipulator. In this case, the plant without input saturation is already nonlinear but is characterized by useful properties (such as feedback linearizability) which provide constructive techniques for high performance nonlinear control laws. When saturation is taken into account, these control laws exhibit a similar behavior to the windup phenomenon widely studied in linear control systems. Indeed, the windup effects on nonlinear saturated control systems is often even worse than the parallel effect in the linear control setting. When dealing with nonlinear plants, we can no longer refer to "desirable linear responses" and the two above mentioned anti-windup requirements need to be rephrased as follows:

1. preserve the unconstrained response arising from the direct interconnection between the nonlinear plant and the nonlinear controller (without saturation) as long as the plant input does not exceed the saturation limits;
2. guarantee as much as possible the recovery of this unconstrained (nonlinear) closed-loop response for all other trajectories.

In this chapter we address the anti-windup design problem for robotic manipulators. In recent years, this problem has been indirectly tackled in the context of anti-windup design for nonlinear plants. In the discrete-time setting, nonlinear anti-windup design techniques have been applied to nonlinear systems in [2,1]. Interesting results related to the nonlinear anti-windup problem can also be found in [12,5], where the attention is restricted to SISO nonlinear plants. MIMO nonlinear plants are considered in [7,6]. However, only local stability results are proven in [6] and restrictions on the local design are necessary in some cases. In [7], the open-loop plant and other subsystems internal to the closed loop are constrained to be asymptotically stable.

Differently from the papers listed above, we explicitly address the problem of anti-windup design for saturated robotic manipulators here, with the goal in mind of guaranteeing high-performance global results. In particular, we improve our work recently appeared in [10], where the ideas of [11] were employed to provide explicit anti-windup constructions for Euler-Lagrange systems.

The goal of this chapter is twofold. The first goal is to clarify the construction suggested in [10] when applied to robotic manipulators (which is the main application field for the theory in [10]). The second and main goal is to revisit and improve the anti-windup laws of [10] to guarantee extreme performance levels on the saturated closed-loop system with anti-windup augmentation. To provide compensation laws that are simple to apply, we explain how the anti-windup gains should be selected and tuned for achieving high performance compensation on generic robot manipulators. Indeed, the parameter tuning boils down to the selection of a proportional and a derivative gain for each degree of freedom of the robotic structure.

The chapter is structured as follows: in Section 2 we describe the anti-windup problem and lay down some useful notation; in Section 3 we first report on the results of [10] and then extend these result to allow for high-performance anti-windup designs; in Section 4 we discuss useful characterizations of the anti-windup performance and, based on these, we provide a simple selection strategy for the anti-windup parameters. Finally, in Section 5 we show the performance of the proposed anti-windup laws on several examples.

2 Problem Data

We will consider in this chapter rigid robot manipulators taking into account the actuator limits affecting their input signals. Given a manipulator belonging to this family, denoting by $q \in I\!\!R^n$ the n joint position variables and by $\dot{q} \in I\!\!R^n$ the corresponding velocity variables, it is well known that the manipulator can be modeled by the following dynamic equations:

$$B(q)\ddot{q} + C(q,\dot{q})\dot{q} + R(q)\dot{q} + h(q) = u_p, \tag{1}$$

where $B(q)$ is the generalized inertia matrix, $C(q,\dot{q})\dot{q}$ represents the generalized centrifugal and Coriolis terms, $h(q)$ is the vector of gravitational forces, the function $R(q)\dot{q}$ represents the friction forces and u_p represents the external forces/torques applied at the robot joints.

The following basic assumption on the regularity of the matrices characterizing (1) will be necessary to prove the main results of this chapter. This assumption derives from standard properties characterizing mechanical systems.

Assumption 1 *The following properties hold:*

1. *the generalized inertia matrix $q \mapsto B(q)$ is continuously differentiable, symmetric and there exist positive numbers λ_M and λ_m such that $\lambda_m I \leq B(q) \leq \lambda_M I$ for all $q \in I\!\!R^n$ (where I denotes the identity);*
2. *the matrix function $(q,\dot{q}) \mapsto C(q,\dot{q})$ is continuous;*
3. *the vector of gravitational forces $q \mapsto h(q)$ is locally Lipschitz;*
4. *the dissipation matrix $q \mapsto R(q)$ is locally Lipschitz and positive semidefinite.*

For the robotic manipulator (1), under Assumption 1, we will assume in this chapter that a (nonlinear) controller, called *unconstrained controller*, henceforth, has been designed such that, when connected in feedback with the robot *without* input saturation, global asymptotic and local exponential stability of the arising closed loop is guaranteed. One such controller is the following feedback linearizing controller with PID action (also known as "computed torque" controller), which is able to induce linear closed-loop behavior (therefore global exponential stability) when saturation is not present and that will be used here to achieve linear decoupled set-point regulation tasks:

$$\begin{aligned}
\dot{x}_c &= q - r \\
u_p &= B(q)\left(-K_p(q-r) - K_d\dot{q} - K_i x_c\right) \\
&\quad + C(q,\dot{q})(\dot{q}) + R(q)(\dot{q}) + h(q),
\end{aligned} \tag{2}$$

where $x_c \in I\!\!R^n$ is the state of the controller and K_p, K_d, K_i are suitable (typically diagonal) square matrices chosen in such a way that the matrix

$$\begin{bmatrix} 0 & I & 0 \\ 0 & 0 & I \\ -K_p & -K_d & -K_i \end{bmatrix}$$

describing the (linear) closed loop (1), (2), is Hurwitz. Based on the value of the reference input $r \in I\!\!R^n$, the controller (2) is able to globally asymptotically stabilize the position $(q, \dot{q}) = (r, 0)$ when interconnected to the robot (1). Note that many alternative controllers could be selected in place of the unconstrained controller (2). Indeed, the only assumption that this needs to satisfy is that it induces global asymptotic and local exponential stability on the closed-loop system with the unconstrained plant. Possible examples are the PD controllers with gravity compensation (nonlinear gravity compensation or constant steady-state gravity compensation). In particular, these last controllers may be more suitable for the set-point regulation tasks that we consider in the rest of this chapter. Nevertheless, we select here a feedback linearizing controller (computed torque) because it better illustrates the desirable local properties of the anti-windup compensation law, where the linear decoupled behavior induced by the feedback linearizing action is preserved whenever the controller output remains within the saturation limits and graceful performance degradation is achieved otherwise.

In this chapter we will characterize the input nonlinearity of (1) as a symmetric decentralized saturation function. This characterization aims at describing the presence of a pool of actuators, one at each joint of the robotic structure, each of them associated with a maximum torque/force effort m_i attainable from the related power unit/motor combination. The saturation function $\mathrm{sat}(\cdot) : I\!\!R^n \to I\!\!R^n$ is therefore defined as

$$\mathrm{sat}(u) = [\,\sigma_1(u_1) \quad \cdots \quad \sigma_n(u_n)\,]^T$$

where

$$\sigma_i(u_i) := \begin{cases} m_i & u_i \geq m_i \\ -m_i & u_i \leq -m_i \\ u_i & -m_i < u_i < m_i. \end{cases} \tag{3}$$

The approach that we propose here could also be applied to non-symmetric saturations, however for simplicity of notation we only consider the symmetric case.

Since the control input of the robotic system (1) is bounded by the presence of the saturation nonlinearity, suitable lower bounds on the saturation levels m_i, $i = 1, \ldots, n$ need to be imposed to guarantee that the actuator has enough power to sustain the robotic structure against the acceleration arising from the gravitational effects. To this aim, we formalize in the following assumption the requirement that the actuators are powerful enough to be able to compensate the gravitational forces in any configuration of the robot (corresponding to a selection of $q \in I\!\!R^n$) with zero velocity.

Assumption 2 *Given the gravitational forces vector $h(\cdot)$ of the robotic system (1) and the saturation limits m_i, $i = 1, \ldots, n$ in (3), the following inequalities hold:*

$$h_{Mi} := \sup_{q \in \mathbb{R}^n} |h(q)| < m_i, \quad i = 1, \cdots, n. \tag{4}$$

The windup problem discussed in the Introduction arises when the controller (2) is no longer interconnected to the plant *without* input saturation but saturation is accounted for in the interconnection. The typical effects of saturation on the closed-loop behavior are to preserve the desirable unconstrained behavior when signals are small enough not to reach the saturation limits and to cause performance and (often) stability loss when signals become large enough so that the saturation enforces modifications at the plant control input.

3   A Nonlinear Anti-Windup Solution

3.1   Prior Work

In this section, we summarize the contribution of [10], when applied to robotic manipulators (which can be described by equations of the type (1)). As shown in Fig. 1, this anti-windup solution corresponds to the insertion of an "anti-windup compensator" as an augmentation to the original control law (2).

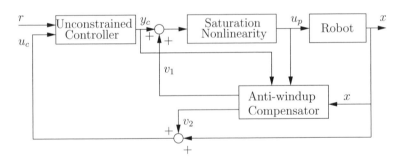

Fig. 1. The anti-windup scheme for robot manipulators.

According to Fig. 1, in the following we will denote by $x := (q, \dot{q}) \in \mathbb{R}^{2n}$ the state of the robot, by $y_c \in \mathbb{R}^n$ the controller output, by $u_p = \text{sat}(u) \in \mathbb{R}^n$ the robot torque/force input and by $u_c = x + v_2 \in \mathbb{R}^{2n}$ the measurement input of the controller. The anti-windup compensator has access to the plant's state and input and to the controller output. The authority of the anti-windup compensator, which allows adding modifications to the original closed loop, consists in two compensation signals v_1 and v_2 which are injected at the controller output and input, respectively. Based on the general approach in [10], when considering robot manipulators, we can provide simplified expressions of the compensation laws implemented in the "anti-windup compensator" block of Fig. 1. In particular, denoting the anti-windup compensator's state by $x_e := (q_e, \dot{q}_e) \in I\!\!R^{2n}$, its dynamics can be written as

$$
\begin{aligned}
\ddot{q}_e = {} & B^{-1}(q)\left(\text{sat}(y_c + v_1) - C(q, \dot{q})\dot{q} - R(q)\dot{q} - h(q)\right) \\
& - B^{-1}(q - q_e)(y_c - C(q - q_e, \dot{q} - \dot{q}_e)(\dot{q} - \dot{q}_e) \\
& \qquad - R(q - q_e)(\dot{q} - \dot{q}_e) - h(q - q_e)).
\end{aligned}
\tag{5}
$$

The anti-windup compensator outputs $v_1 \in I\!\!R^n$ and $v_2 \in I\!\!R^{2n}$ correspond to

$$
\begin{aligned}
v_1 &= \beta(x, x_e) \\
v_2 &= -x_e = -(q_e, \dot{q}_e),
\end{aligned}
\tag{6}
$$

where $\beta(\cdot, \cdot) : I\!\!R^{2n} \times I\!\!R^{2n} \to I\!\!R^n$ is given by

$$
v_1 = \beta(x, x_e) := h(q) - h(q - q_e) - K_g \text{sat}(K_g^{-1} q_e) - K_0 \dot{q}_e.
\tag{7}
$$

The two matrices K_0 and K_g are positive definite diagonal and they represent the "tuning" parameters of the anti-windup law. The diagonal elements $\kappa_{gi}, i = 1, \ldots, n$ of K_g need to satisfy the following constraints:

$$
h_{Mi} + \kappa_{gi} m_i < m_i, \quad i = 1, \ldots, n.
\tag{8}
$$

Note that by definition of h_{Mi} in (4), if Assumption 2 holds, there always exists a positive definite diagonal matrix K_g fulfilling the constraints (8).

Given the construction above, we report in the following, for the sake of clarity, the complete control scheme with anti-windup compensation built on top of the computed torque controller (2):

$$
\begin{aligned}
\dot{x}_c &= q - q_e - r \\
y_c &= B(q - q_e)\left(-K_p(q - q_e - r) - K_d(\dot{q} - \dot{q}_e) - K_i x_c\right) \\
& \quad + C(q - q_e, \dot{q} - \dot{q}_e)(\dot{q} - \dot{q}_e) + R(q - q_e)(\dot{q} - \dot{q}_e) + h(q - q_e), \\
\ddot{q}_e &= B^{-1}(q)\left(u_p - C(q, \dot{q})\dot{q} - R(q)\dot{q} - h(q)\right) \\
& \quad - B^{-1}(q - q_e)(y_c - C(q - q_e, \dot{q} - \dot{q}_e)(\dot{q} - \dot{q}_e) \\
& \qquad\qquad - R(q - q_e)(\dot{q} - \dot{q}_e) - h(q - q_e)) \\
u_p &= \text{sat}(y_c + v_1),
\end{aligned}
\tag{9}
$$

where v_1 is selected as in (7). If an alternative unconstrained controller was used, then the first two equations above should be replaced by its dynamics.

The main result of [10] establishes useful properties of the trajectories of the *anti-windup closed-loop system* (1), (9), (7) (whose state will be denoted by (x, x_c, x_e)) when compared to the (ideal) trajectories of the unconstrained closed-loop system (1), (2) (whose state will be denoted using the subscript "ℓ", namely $(x_\ell, x_{c\ell})$). This is formalized in the following theorem (reported here without proof).

Theorem 1. *[10] Suppose that Assumptions 1 and 2 hold and the parameters of the compensation law (7) satisfy (8). Given a constant reference signal r, denote by $(x_\ell(t), x_{c\ell}(t))$ the response of the unconstrained closed-loop system (1), (2) starting from the initial conditions $((x_\ell(0), x_{c\ell}(0))$. Denote also by $u_\ell(t)$ the corresponding controller output. Then the anti-windup closed-loop system (1), (9), (7) is such that*

1. *if* $u_\ell(t) = \mathrm{sat}(u_\ell(t))$ *for all times and* $(x(0), x_c(0), x_e(0)) = (x_\ell(0), x_{c\ell}(0), 0)$, *then* $(x(t), x_c(t), x_e(t)) = (x_\ell(t), x_{c\ell}(t), 0)$ *for all times, namely the unconstrained response is retained;*
2. *defining* $x^* := (r, 0)$, *there exists a vector* $x_c^* \in \mathbb{R}^n$ *such that the point* $(x^*, x_c^*, 0)$ *is globally asymptotically stable and locally exponentially stable.*

Theorem 1 establishes two important properties of the anti-windup closed-loop system (1), (9), (7). The first one corresponds to the key constraint of anti-windup construction discussed in the Introduction: the anti-windup compensation must preserve the local response of the original (unconstrained) closed loop whenever the saturation limits are not exceeded by the unconstrained trajectory. The second property states that the closed loop with anti-windup augmentation is globally asymptotically (and locally exponentially) stable, thus the instability effects often experienced when control laws such as (2) reach the saturation limits (see Section 5 for a notable example of this phenomenon) are eliminated by the proposed anti-windup augmentation strategy.

3.2   A Generalized Result

If on one hand the result of the previous section guarantees important properties of our anti-windup augmentation scheme, very little is established about the transient response of the anti-windup closed-loop system after the saturation limits are reached by the actuators: in this case, the only property guaranteed by Theorem 1 (in particular, by item 2) is that the closed-loop trajectories converge to the desired equilibrium point where $q = r$ and $\dot{q} = 0$. Nothing can be concluded, however, about the transient behavior of these trajectories. To allow for high performance selections of the anti-windup compensator parameters (the selection method will be clarified in the following section), we introduce in this section an extension of the anti-windup law of [10] summarized above. In particular, we propose the following generalization of the selection for v_1 in (7):

$$v_1 = \mathrm{sat}(y_c) - y_c + h(q) - h(q - q_e) - K_g\mathrm{sat}(K_g^{-1}K_q q_e) - K_{qd}(q_e, \dot{q}_e)\dot{q}_e, \quad (10)$$

where K_g is a diagonal matrix whose elements still satisfy the constraints (8), K_q is a diagonal positive definite matrix and $K_{qd}(\cdot, \cdot)$ is a decentralized diagonal matrix function, constant in a neighborhood of the origin, with diagonal elements $\kappa_{qdi}(\cdot, \cdot)$, $i = 1, \ldots, n$ such that the maps $(q_{ei}, \dot{q}_{ei}) \mapsto \kappa_{qdi}(q_{ei}, \dot{q}_{ei})\dot{q}_{ei}$, $i = 1, \ldots, n$ are scalar locally Lipschitz functions and such that there exists a positive constant $\underline{\kappa}_{qd}$ which satisfies

$$\kappa_{qdi}(q_{ei}, \dot{q}_{ei}) \geq \underline{\kappa}_{qd}, \quad \forall q_{ei}, \dot{q}_{ei} \in \mathbb{R}, \forall i \in \{1, \ldots, n\}. \quad (11)$$

By suitably generalizing the proof of the main result of [10] (corresponding to Theorem 1 above), the following parallel result can be established for the generalized anti-windup closed-loop system arising from the interconnection between (1), (9) and the new compensation signal (10). The proof of the following theorem is omitted because of its similarity with Theorem 1 and due to space constraints.

Theorem 2. *Suppose that Assumptions 1 and 2 hold and the parameters of the compensation law (10) satisfy (8) and (11).*

Then the anti-windup closed-loop system (1), (9), (10) satisfies the same anti-windup properties established in Theorem 1.

Note that the compensation law (10) is a generalization of (7). This generalization allows for significant performance improvements as compared to the results reported in [10] (where the compensation law (7) was employed). To this aim, in the next section we will first characterize mathematically the performance of the anti-windup compensation scheme and then describe suitable selections of the parameters K_g, K_q and $K_{qd}(\cdot, \cdot)$ in (10) that are especially effective at guaranteeing high performance compensation.

4   Measuring and Improving the Anti-Windup Performance

Following the anti-windup qualitative goal of recovering as much as possible "what the response without input saturation would be", the quality of the closed-loop response can be measured in terms of the deviation of the actual plant trajectory x from the corresponding (ideal) unconstrained plant trajectory x_ℓ. In particular, we are interested in the size of the signal $x(t) - x_\ell(t)$ for all positive times (and item 2 of Theorem 1 guarantees that, for any constant reference r, $x(t) - x_\ell(t)$ converge to zero because both these signals converge to the equilibrium $(r, 0)$).

While item 1 of Theorem 1 guarantees that $x(t) - x_\ell(t)$ is identically zero when $u_\ell(\cdot)$ never exceeds the saturation limits, no information about the transient behavior of $x(t) - x_\ell(t)$ is available from the theorem for all other trajectories. On the other hand, based on continuity of trajectories with respect to initial conditions on compact time intervals (this is a standard result of nonlinear systems analysis) and on the global asymptotic stability property of item 1, it is reasonable to expect that unconstrained trajectories corresponding to control inputs u_ℓ that spend little time (and little energy) outside the saturation limits will correspond to trajectories of the anti-windup closed-loop system such that $x(t) - x_\ell(t)$ is very small (in some sense).

For all the remaining trajectories, not much can be concluded about their transient behavior from Theorem 1. For these cases, the following result is a good starting point to monitor and, possibly, make small the size of $x(t) - x_\ell(t)$.

Theorem 3. *Regardless of the selection of v_1 in (2), given any reference signal $r(t), t \geq 0$, denote by $(x_\ell(t), x_{c\ell}(t))$ the response of the unconstrained closed-loop system (1), (2) starting from the initial conditions $((x_\ell(0), x_{c\ell}(0))$ and denote by $(x(t), x_c(t), x_e(t))$ the response of the anti-windup closed-loop system (1), (9), (7) starting from the initial conditions $(x(0), x_c(0), x_e(0)) = (x_\ell(0), x_{c\ell}(0), 0)$. Then*

$$x_e(t) = x_\ell(t) - x(t), \quad \forall t \geq 0.$$

Proof. Consider the closed loop (1), (9) and perform the change of coordinates $(x, x_c, x_e) \rightarrow (\tilde{x}, x_c, x_e)$, where $\tilde{x} := x - x_e$. Then, defining $(\tilde{q}, \dot{\tilde{q}}) := \tilde{x}$, after some computation, the following equations are obtained:

$$\begin{cases} \ddot{\tilde{q}} = -B^{-1}(\tilde{q}) \left(C(\tilde{q}, \dot{\tilde{q}})\dot{\tilde{q}} + R(\tilde{q})\dot{\tilde{q}} + h(\tilde{q}) - y_c \right) \\ \dot{x}_c = \tilde{q} - r \\ y_c = B(\tilde{q}) \left(-K_p(\tilde{q} - r) - K_d\dot{\tilde{q}} - K_i x_c \right) + C(\tilde{q}, \dot{\tilde{q}})\dot{\tilde{q}} + R(\tilde{q})\dot{\tilde{q}} + h(\tilde{q}), \end{cases} \quad (12)$$

$$\begin{cases} \ddot{q}_e = B^{-1}(\tilde{q} + q_e)(u_p - C(\tilde{q} + q_e, \dot{\tilde{q}} + \dot{q}_e)(\dot{\tilde{q}} + \dot{q}_e) \\ \qquad\qquad - R(\tilde{q} + q_e)(\dot{\tilde{q}} + \dot{q}_e) - h(\tilde{q} + q_e)) \\ \qquad + B^{-1}(\tilde{q}) \left(C(\tilde{q}, \dot{\tilde{q}})\dot{\tilde{q}} + R(\tilde{q})\dot{\tilde{q}} + h(\tilde{q}) - y_c \right) \\ u_p = \text{sat}(y_c + v_1). \end{cases} \quad (13)$$

The representation (12), (13) for the anti-windup closed-loop system is the cascade of two subsystems. The first one (corresponding to (12)) of coordinates $(\tilde{x}, x_c)$ driving a second one (corresponding to (13)) of coordinates x_e. Note that the dynamics (12) of the first subsystem are coincident with the unconstrained dynamics (1), (2) and that, since $x_e(0) = 0$, then $(\tilde{x}(0), x_c(0)) = (x(0), x_c(0))$. Consequently, since the dynamics and the initial conditions are the same, $(\tilde{x}(t), x_c(t)) = (x_\ell(t), x_{c\ell}(t))$ for all positive times. Therefore, by definition, $x_\ell(t) = \tilde{x}(t) = x(t) - x_e(t)$ for all positive times and the result follows.

From a performance perspective, the relevance of Theorem 3 stands in the fact that it clarifies the impact of the selection of v_1 on the error variables $x_e = x - \tilde{x}$. By virtue of the cascade structure (12), (13) pointed out in the proof of Theorem 3, we can focus on the second dynamics (13) to study selections of v_1 of the type (10) that are particularly effective at keeping q_e small, so that the actual trajectory q is as close as possible to the (ideal) unconstrained trajectory $\tilde{q}$. Note, however, that the global asymptotic (and local exponential) stability of (13) is already assured by Theorem 2 for all selections of the parameters that satisfy (8) and (11), so we can disregard the stability property (which has already been addressed and proven) and concentrate on performance.

A first thing to point out is the fact that, according to the second equation in (13), the term y_c acts like a disturbance for the dynamics q_e. This motivates the term $\text{sat}(y_c) - y_c$ in equation (10) which alone leads to highly improved responses (as compared to (7)) in the first instants of the closed-loop response. Indeed, especially in aggressive control systems, y_c often presents very large peaks that result in undesired undershoots at the beginning of the anti-windup closed-loop response. Adding this extra term transforms the disturbance from y_c into $\text{sat}(y_c)$, thus reducing significantly its negative effects. [1]

To understand the impact of the selection (10) on the error dynamics (13), it is useful to substitute v_1 and u_p in the first equation of (13). We are especially interested in the dynamics of q_e associated with times where the plant input is not anymore

---

[1] One may think that the best strategy is to eliminate completely y_c. However, it would not be possible to guarantee item 1 of Theorem 2 in that case.

saturated, so that full authority is available for the signal v_1 to suitably drive the state x_e. Therefore, substituting $u_p = y_c + v_1$ in the first equation of (13) we get (recall that $\tilde{q} = q - q_e$):

$$
\begin{aligned}
\ddot{q}_e = {} & B^{-1}(\tilde{q} + q_e)\left(-K_g\mathrm{sat}(K_g^{-1}K_q q_e) - K_{qd}(q_e)\dot{q}_e\right) \\
& + B^{-1}(\tilde{q} + q_e)(\mathrm{sat}(y_c) - C(\tilde{q} + q_e, \dot{\tilde{q}} + \dot{q}_e)(\dot{\tilde{q}} + \dot{q}_e) \\
& - R(\tilde{q} + q_e)(\dot{\tilde{q}} + \dot{q}_e) - h(\tilde{q})) - \ddot{\tilde{q}}
\end{aligned}
$$

Interestingly, it follows that when $(q_e, \dot{q}_e)$ is small and $\mathrm{sat}(y_c) = y_c$, by continuity, the second line of the above equation is almost zero and if the saturation on the first line is not active we get

$$
B(q)\ddot{q}_e \approx -K_g\mathrm{sat}(K_g^{-1}K_q q_e) - K_{qd}(q_e, \dot{q}_e)\dot{q}_e, \tag{14}
$$

which describes a dynamic system close to a double integrator controlled by a saturated proportional action and by a derivative action, whose gains are associated with the design parameters K_q and $K_{qd}(\cdot, \cdot)$ (recall that $K_{qd}(\cdot, \cdot)$ is diagonal and strictly positive for all values of its arguments, by construction).

Let us denote by $\gamma_E(q_e)$ the equivalent gain associated with the saturation of the proportional action, namely $\gamma_E(\cdot)$ is a diagonal matrix function which satisfies $K_g\mathrm{sat}(K_g^{-1}K_q q_e) = \gamma_E(q_e)K_q q_e$. Let us also denote by $D(q_e, \dot{q}_e)$ a diagonal matrix whose diagonal elements $d_i, i = 1, \ldots, n$ are selected as follows:

$$
d_i = \begin{cases} 1, & \text{if } q_{ei}\dot{q}_{ei} \geq 0 \\ 0, & \text{otherwise,} \end{cases}
$$

where q_{ei} and $\dot{q}_{ei}$, $i = 1, \ldots, n$ are the components of q_e and $\dot{q}_e$, respectively. Then given a positive definite diagonal matrix K_0, we select the function $K_{qd}(\cdot, \cdot)$ so that

$$
K_{qd}(q_e, \dot{q}_e)\dot{q}_e := \left(\left(1 - D(q_e, \dot{q}_e)\right)\gamma_E(q_e) + D(q_e, \dot{q}_e)\right)K_0\dot{q}_e. \tag{15}
$$

The selection (15) is easily explained by first noting that, with respect to each component of q_e (and $\dot{q}_e$), when q_{ei} and $\dot{q}_{ei}$ have the same sign, so that both the proportional and the derivative term have the same sign in (14), then $K_{qd}(q_e, \dot{q}_e) = K_0$, regardless of the size of both q_{ei} and $\dot{q}_{ei}$. However, if q_{ei} and $\dot{q}_{ei}$ have opposite signs, so that the derivative term in (14) is exerting a breaking force/torque, then $K_{qd}(q_e, \dot{q}_e) = \gamma_E(q_e)K_0$, so that such a breaking action is modulated by the depth into saturation of the proportional element. [2] This modulating action leads to significant performance improvement when q_e is very large and the saturated proportional term in (14) becomes too small as compared to the breaking action arising from the derivative term. Note that with the selection (10), (15), if q_e is small enough so that the second saturation function in (10) is not active, since $\gamma_E(q_e) = I$, the approximate dynamics (14) transform into the simple dynamics

$$
B(q)\ddot{q}_e \approx -K_q q_e - K_0\dot{q}_e. \tag{16}
$$

---

[2] Note that since the operating region of the robot is bounded, by the closed-loop stability established in Theorem 2, also q_e (consequently, $\gamma_E(q_e)$) is bounded, and thus the selection (15) satisfies the constraint (11).

Based on this property, it is straightforward to show that the selection (15) is Lipschitz. Moreover, equation (16) suggest that the diagonal elements of K_q and K_0 should be selected in an almost decoupled way ("almost" because of the presence of $B(q)$), with the goal of improving the performance at each joint, following a selection approach similar to the heuristic approach for the selection of linear PD gains.

Summarizing the above, a successful strategy for the selection of v_1 is (10), (15), whose design parameters are three positive definite diagonal matrices K_g, K_q, K_0. The first parameter, K_g, should always be chosen as large as possible within the design constraints (8) to maximize the authority of the proportional gain in the compensation law (note that $K_g < I$ by definition). The parameters K_q and K_0 should be tuned with the goal of improving the transients at each joint following a quasi decoupled PD tuning strategy.

5 Anti-Windup Construction Examples

In this section we will consider three simulation examples to demonstrate the proposed anti-windup construction. The first example will be useful to understand the implications of the selection (15) on a linear decoupled mechanical system. The second example shows the performance of the proposed construction on a simple nonlinear robot arm. Finally, the construction is applied to the same model used in [10], showing the dramatic performance improvement arising from the design method proposed here.

To simplify the notational burden, throughout this section we will often denote the components of a vector w by suitably adding subscripts to the vector name (so that, e.g., $y_{c_1}, \ldots, y_{c_n}$ may denote the components of the vector y_c). Moreover, given two vectors a, b, we will use (a, b) to denote the vector $[a^T \ b^T]^T$.

5.1 Planar Positioning System

In this example, we will show the impact of the anti-windup law on a linear and decoupled robot, in which the input constraints lead to severe performance loss in the saturated closed loop without anti-windup.

System model The positioning system is a two-link robot, with two prismatic joints. The model is very simple: the robot is not subject to gravitational force, the generalized inertia matrix is linear and decoupled and so is the matrix containing the friction terms. A schematic diagram of the planar positioning system is reported in Fig. 2.

The system model is expressed by the following equations

$$
\begin{aligned}
(M_1 + M_2)\ddot{q}_1 + \rho_1 \dot{q}_1 &= u_{p_1} \\
M_2 \ddot{q}_2 + \rho_2 \dot{q}_2 &= u_{p_2}
\end{aligned}
\tag{17}
$$

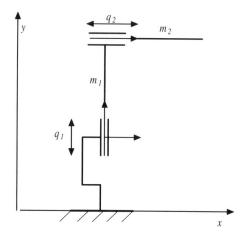

Fig. 2. The planar positioning system.

where M_i is the total mass of each link (including the actuators' mass), ρ_i is the friction coefficient of the i-th link, and u_{pi} is the actuator force.

The values of parameters are reported in Tab. 1, where m_i is the saturation level of each actuator.

Table 1. Parameters of the planar positioning system.

Link	M_i [kg]	m_i [N]	ρ_i [Kg/s]
1	3	40	2
2	2	40	1

Unconstrained controller design The unconstrained control system is a "computed torque" controller, which is able to induce global exponential stability when saturation does not occur. The performance of the unconstrained controller is obtained choosing suitable values for the diagonal matrices K_p, K_d, K_i. The equations of the unconstrained controller are:

$$\dot{x}_{c_1} = \tilde{q}_1 - r_1$$
$$\dot{x}_{c_2} = \tilde{q}_2 - r_2$$
$$y_{c_1} = \left(M_1 + M_2\right)\left(-k_{p_1}(\tilde{q}_1 - r_1) - k_{d_1}\dot{\tilde{q}}_1 - k_{i_1}x_{c_1}\right) + \rho_1\dot{\tilde{q}}_1$$
$$y_{c_2} = M_2\left(-k_{p_2}(\tilde{q}_2 - r_2) - k_{d_2}\dot{\tilde{q}}_2 - k_{i_2}x_{c_2}\right) + \rho_2\dot{\tilde{q}}_2$$

where $(\tilde{q}, \dot{\tilde{q}}) = ([\tilde{q}_1\ \tilde{q}_2]^T, [\dot{\tilde{q}}_1\ \dot{\tilde{q}}_2]^T)$ represents the controller input, so that the unconstrained interconnection corresponds to

$$(\tilde{q}, \dot{\tilde{q}}) = (q, \dot{q}), \qquad u_p = y_c, \tag{18}$$

the saturated interconnection (without anti-windup) corresponds to

$$(\tilde{q}, \dot{\tilde{q}}) = (q, \dot{q}), \qquad u_p = \text{sat}(y_c), \tag{19}$$

and the anti-windup interconnection corresponds to

$$(\tilde{q}, \dot{\tilde{q}}) = (q - q_e, \dot{q} - \dot{q}_e), \qquad u_p = \text{sat}(y_c + v_1), \tag{20}$$

where $(q_e, \dot{q}_e)$ is the anti-windup compensator's state. The proportional, integral and derivative gains of the unconstrained controller have been selected as follows:

$$K_p = \text{diag}(360, 360)$$
$$K_d = \text{diag}(30, 30)$$
$$K_i = \text{diag}(8, 8).$$

Anti-windup design and tuning The anti-windup construction consists in writing the anti-windup compensator dynamics and in choosing the parameters of the control law (10). By substituting equations (17) in (5) we obtain:

$$\ddot{q}_{e_1} = \tfrac{1}{M_1+M_2}\left(u_{p_1} - \rho_1 \dot{q}_1\right) + \tfrac{1}{M_1+M_2}\left(\rho_1(\dot{q}_1 - \dot{q}_{e_1}) - y_{c_1}\right)$$
$$\ddot{q}_{e_2} = \tfrac{1}{M_2}\left(u_{p_2} - \rho_2 \dot{q}_2\right) + \tfrac{1}{M_2}\left(\rho_2(\dot{q}_2 - \dot{q}_{e_1}) - y_{c_2}\right)$$

where $\begin{bmatrix} u_{p_1} & u_{p_2} \end{bmatrix}^T$ contains the force input, which corresponds to:

$$u_{p_1} = \sigma_1\left(\sigma_1(y_{c_1}) - k_{g_1}\sigma_1\left(\tfrac{k_{q_1} q_{e_1}}{k_{g_1}}\right) - k_{qd_1}(q_e, \dot{q}_e)\dot{q}_{e_1}\right)$$
$$u_{p_2} = \sigma_2\left(\sigma_2(y_{c_2}) - k_{g_2}\sigma_1\left(\tfrac{k_{q_2} q_{e_2}}{k_{g_2}}\right) - k_{qd_2}(q_e, \dot{q}_e)\dot{q}_{e_2}\right),$$

where $k_{qd_1}(\cdot, \cdot)$ and $k_{qd_2}(\cdot, \cdot)$ are the diagonal elements of the matrix function $K_{qd}(\cdot, \cdot)$ defined in (15).

As for the selection of the diagonal matrix gains K_g, K_q and K_0, since there is no gravity effect on this model, we can select K_g arbitrarily close to the identity, e.g.,

$$K_g = \text{diag}(0.99, 0.99),$$

which satisfies the constraint (8). The remaining matrix gains K_q and K_0 should be selected with the goal of improving the performance of the anti-windup law during the transient response. For each entry $i = 1, 2$, on the diagonal of K_q and K_0, we select the proportional term K_{qi} to guarantee a fast enough convergence of the related component of q_e to zero (namely, by Theorem 3, a fast enough convergence of q to the unconstrained response q_ℓ), and the derivative term K_{0i} to enforce suitable damping on the terminal part of the trajectory, thus avoiding undesirable oscillations of the anti-windup closed-loop response. Following this approach, the parameters are easily tuned as:

$$K_0 = \text{diag}(650, 250)$$
$$K_q = \text{diag}(2600, 1600).$$

Simulation results We test by simulation our construction by selecting the reference signal as the following step input:

$$r = (2.5, 2) \ [\text{m}], \tag{21}$$

and initializing both the plant and the controller states at zero.

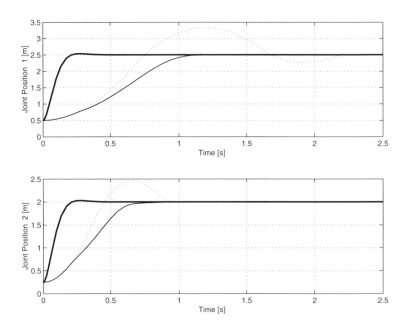

Fig. 3. Planar positioning system: output responses to the reference (21) of the following closed-loop systems: unconstrained (bold solid), saturated without anti-windup (dotted), and anti-windup (thin solid).

The corresponding simulations are reported in Figs. 3 and 4. In Fig. 3, the bold curves represent the unconstrained output responses. The output responses of the anti-windup closed-loop system (thin solid) reach the reference positions in less than one second, eliminating the undesired oscillations exhibited by the saturated closed-loop system without anti-windup (dotted). Figure 4 represents the plant input responses for the same three closed-loop systems. Note that the anti-windup action exploits the full actuators power available to allow for the fast output responses of Fig. 3. This fact becomes evident when noticing that the plant input signals become saturated both during the acceleration and during the deceleration phases.

5.2   Two-Link Planar Robot

In this example, we consider the planar two-link robot arm represented in Fig. 5, displaced on a vertical plane so that the gravitational vector is oriented as shown in

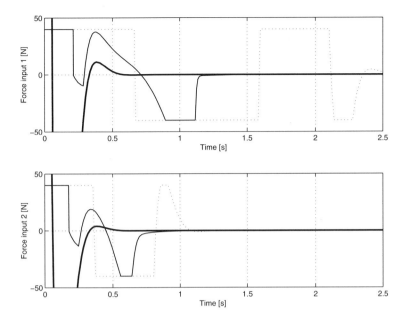

Fig. 4. Planar positioning system: input responses to the reference (21) of the following closed-loop systems: unconstrained (bold solid), saturated without anti-windup (dotted), and anti-windup (thin solid).

the figure. Contrary to the previous example, the robot dynamics is nonlinear and not decoupled, and the gravitational acceleration affects both links. The aim of this example is to show the quasi-decoupled performance of the anti-windup closed-loop system in the presence of input saturation and to illustrate the easy selection of the anti-windup parameters.

System model The planar robot is a two-link robot arm, with two rotational joints, subject to the gravitational force. A schematic representation of the robot is shown in Fig. 5.

According to the notation used in equation (1), we report the generalized inertia matrix $B(q)$, the matrix $C(q, \dot{q})$ containing the centrifugal and Coriolis terms and the gravitational vector $G(q)$. For simplicity, we select the friction forces to be zero. The generalized inertia matrix $B(q)$ corresponds to:

$$B(q) = \begin{bmatrix} b_{11} & b_{12} \\ b_{12} & b_{22} \end{bmatrix}$$

with

$$b_{11} = I_1 + M_1 l_1^2 + I_2 + M_2(a_1^2 + l_2^2 + 2a_1 l_2 \cos(q_2))$$
$$b_{12} = I_2 + M_2(l_2^2 + a_1 l_2 \cos(q_2))$$
$$b_{22} = I_2 + M_2 l_2^2$$

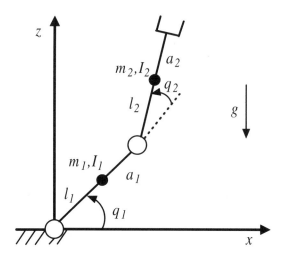

Fig. 5. The two-link planar robot.

where $q = [\,q_1 \quad q_2\,]^T$ contains the two joint variables, (M_1, M_2) are the total masses of the two links (including the actuators' masses), (a_1, a_2) represent the link lengths, (l_1, l_2) represent the distances of the center of mass of each link from the preceding joint, and (I_1, I_2) represent the rotational inertias at the two joints. The matrix $C(q, \dot{q})$ can be written as follows:

$$C(q, \dot{q}) = \begin{bmatrix} -M_2 a_1 l_2 \sin(q_2)\dot{q}_2 & -M_2 a_1 l_2 \sin(q_2)(\dot{q}_1 + \dot{q}_2) \\ M_2 a_1 l_2 \sin(q_2)\dot{q}_1 & 0 \end{bmatrix}.$$

The gravitational vector is

$$h(q) = \big[\, g(M_1 l_1 + M_2 a_1)\cos(q_1) + g M_2 l_2 \cos(q_1 + q_2) \ \ g M_2 l_2 \cos(q_1 + q_2) \,\big]^T$$

where g is the gravitational acceleration value. The physical parameters have been selected as shown in Tab. 2, where (m_1, m_2) represent the saturation levels for the torques exerted at the two joints.

Table 2. Parameters of the two-link planar robot.

Link	l_i [m]	M_i [kg]	I_i [kgm^2]	a_i [m]	m_i [Nm]
1	0.5	6	0.2	1	138
2	0.25	5	0.1	0.5	40

Unconstrained controller design The unconstrained controller is once again selected as a computed torque controller, which induces linear and decoupled closed-loop behavior before saturation is activated. The corresponding equations are:

$$\dot{x}_{c_1} = \tilde{q}_1 - r_1$$
$$\dot{x}_{c_2} = \tilde{q}_2 - r_2$$
$$y_{c_1} = \left(I_1 + M_1 l_1^2 + I_2 + M_2(a_1^2 + l_2^2 + 2a_1 l_2 \cos(\tilde{q}_2))\right)\cdot$$
$$\left(-k_{p_1}(\tilde{q}_1 - r_1) - k_{d_1}\dot{\tilde{q}}_1 - k_{i_1} x_{c_1}\right)$$
$$+\left(I_2 + M_2(l_2^2 + a_1 l_2 \cos(\tilde{q}_2))\right)\left(-k_{p_2}(\tilde{q}_2 - r_2) - k_{d_2}\dot{\tilde{q}}_2 - k_{i_2} x_{c_2}\right)$$
$$-2M_2 a_1 l_2 \dot{\tilde{q}}_1 \dot{\tilde{q}}_2 \sin(\tilde{q}_2) - M_2 a_1 l_2 \dot{\tilde{q}}_2^2 \sin(\tilde{q}_2)$$
$$+g(M_1 l_1 + M_2 a_1)\cos(\tilde{q}_1) + g M_2 l_2 \cos(\tilde{q}_1 + \tilde{q}_2)$$
$$y_{c_2} = \left(I_2 + M_2(l_2^2 + a_1 l_2 \cos(\tilde{q}_2))\right)\left(-k_{p_1}(\tilde{q}_1 - r_1) - k_{d_1}\dot{\tilde{q}}_1 - k_{i_1} x_{c_1}\right)$$
$$+\left(I_2 + M_2 l_2^2\right)\left(-k_{p_2}(\tilde{q}_2 - r_2) - k_{d_2}\dot{\tilde{q}}_2 - k_{i_2} x_{c_2}\right)$$
$$+M_2 a_1 l_2 \dot{\tilde{q}}_1^2 \sin(q_2) + g M_2 l_2 \cos(\tilde{q}_1 + \tilde{q}_2)$$

where $(\tilde{q}, \dot{\tilde{q}})$ represents the controller input, so that, similar to the previous example, the unconstrained interconnection corresponds to (18), the saturated interconnection (without anti-windup) corresponds to (19) and the anti-windup interconnection corresponds to (20). The proportional, integral and derivative gains of the unconstrained controller have been selected as follows:

$$K_p = \text{diag}(240, 255)$$
$$K_d = \text{diag}(45, 50)$$
$$K_i = \text{diag}(4, 4).$$

Anti-windup design and tuning Similar to the previous example, based on (5), the anti-windup compensator dynamics can be written as:

$$\ddot{q}_e = B^{-1}(q)(u_p - C(q, \dot{q})(q, \dot{q}) - G(q))$$
$$+B^{-1}(q - q_e)(C(q - q_e, \dot{q} - \dot{q}_e)(q - q_e, \dot{q} - \dot{q}_e) + G(q - q_e) - y_c)$$
$$u_p = \text{sat}(y_c + v_1)$$

where y_c is the unconstrained controller output and v_1 is the anti-windup control law expressed by

$$v_1 = \left(\sigma_1(y_{c_1}) - y_{c_1}\right) + g(M_1 l_1 + M_2 a_1)\cos(q_1) + g M_2 l_2 \cos(q_1 + q_2)$$
$$-g(M_1 l_1 + M_2 a_1)\cos(q_1 - q_{e_1}) - g M_2 l_2 \cos(q_1 - q_{e_1} + q_2 - q_{e_2})$$
$$-k_{g_1}\sigma_1\left(\frac{k_{q_1} q_{e_1}}{k_{g_1}}\right) - k_{qd_1}(q_e, \dot{q}_e)\dot{q}_{e_1}$$
$$v_2 = \left(\sigma_2(y_{c_2}) - y_{c_2}\right) + g M_2 l_2 \cos(q_1 + q_2) - g M_2 l_2 \cos(q_1 - q_{e_1} + q_2 - q_{e_2})$$
$$-k_{g_2}\sigma_1\left(\frac{k_{q_2} q_{e_2}}{k_{g_2}}\right) - k_{qd_2}(q_e, \dot{q}_e)\dot{q}_{e_2},$$

where $k_{qd_1}(\cdot, \cdot)$ and $k_{qd_2}(\cdot, \cdot)$ are the diagonal elements of the matrix function $K_{qd}(\cdot, \cdot)$ defined in (15).

The diagonal elements of the matrix K_g have been chosen to satisfy he constraint (8) as follows:

$$K_g = \text{diag}(0.29, 0.64).$$

The diagonal elements of the matrices K_q, K_0 have been selected once again following the approach outlined in Section 4. The resulting matrices are:

$$K_0 = \text{diag}(150, 400)$$
$$K_q = \text{diag}(400, 400).$$

Simulation results Once again, the reference signal has been selected as a step input assuming the following values:

$$r = (90, -45) \text{ [deg]}, \tag{22}$$

and both the plant and the controller states have been initialized at zero.

The corresponding simulations are reported in Figs. 6 and 7. Once again, the bold curves represent the unconstrained responses, the dotted curves represent the saturated responses (without anti-windup) and the thin solid curves represent the anti-windup responses. Observe that the undesired undershoot presented by the saturated response is completely eliminated by the anti-windup action. Moreover, the anti-windup compensation is able to almost fully preserve the linear performance at the second joint. This is not the case in the first joint response, which exhibits an inevitable response delay due to the input limitation.

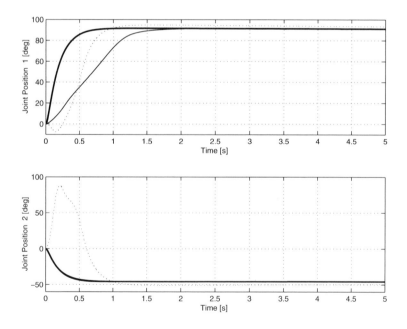

Fig. 6. Two link planar robot: output responses to the reference (22) of the following closed-loop systems: unconstrained (bold solid), saturated without anti-windup (dotted), and anti-windup (thin solid).

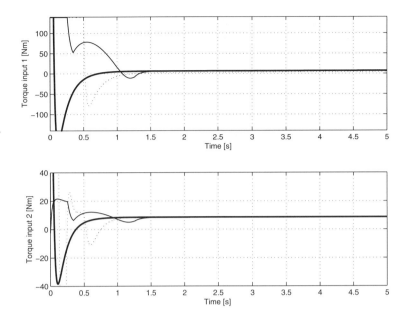

Fig. 7. Two link planar robot: input responses to the reference (22) of the following closed-loop systems: unconstrained (bold solid), saturated without anti-windup (dotted), and anti-windup (thin solid).

From the input responses reported in Fig. 7, it appears that the anti-windup compensator makes large use of the available input effort. Nevertheless, the input signal does not reach saturation other than a short time interval during the first 200 milliseconds. This suggests that increasing the anti-windup gains K_q and K_0 may lead to a faster response. Additional simulations, which are not reported here, confirm that increasing the anti-windup gains to the values $K_0 = \mathrm{diag}(600, 400)$, $K_q = \mathrm{diag}(4000, 400)$ allows reducing the recovery transient on the first joint from approximately 1.5 seconds to 1 second (the response on the second joint remains the same).

5.3   SCARA Robot

In [10], the effectiveness of the proposed anti-windup law has been tested on a SCARA robot (Selective Compliance Assembly Robot Arm). We use here the same example to emphasize the performance improvement that can be guaranteed when employing the improved anti-windup law given by (10), (15).

System model   The SCARA robot has four links. The first two links correspond to a planar robot on the horizontal plane. The third link corresponds to a prismatic joint imposing the tilt of the end effector on the working surface and the last joint is a rotational joint corresponding to the end effector orientation with respect to the vertical rotation axis. According to the notation in (1), we report in the following the matrices associated to the robot model. The generalized inertia matrix $B(q)$ is

$$B(q) = \begin{bmatrix} b_{11} & b_{12} & b_{13} & b_{14} \\ b_{12} & b_{22} & b_{23} & b_{24} \\ b_{13} & b_{23} & b_{33} & b_{34} \\ b_{14} & b_{24} & b_{34} & b_{44} \end{bmatrix}$$

with

$$b_{11} = I_1 + I_2 + I_3 + I_4 + M_1 l_{c_1}^2 + M_2 \left(l_1^2 + l_{c_2}^2 + 2l_{c_2}l_1\cos(q_2)\right)$$
$$+ (M_3 + M_4)\left(l_1^2 + l_2^2 + 2l_1l_2\cos(q_2)\right)$$
$$b_{12} = I_2 + I_3 + I_4 + M_2(l_{c_2}^2 + l_1 l_{c_2}\cos(q_2)) + (M_3 + M_4)\left(l_1^2 + l_2^2 + l_1 l_2\cos(q_2)\right)$$
$$b_{13} = 0$$
$$b_{14} = -I_4$$
$$b_{22} = I_2 + I_3 + I_4 + M_2 l_{c_2}^2 + M_3 l_2^2 + M_4 l_2^2$$
$$b_{23} = 0$$
$$b_{24} = -I_4$$
$$b_{33} = M_3 + M_4$$
$$b_{34} = 0$$
$$b_{44} = I_4$$

where l_i is the length of the i-th link, l_{c_i} represents the distance between the center of gravity of each link and the center of the preceding joint, M_i is the total mass of the i-th link (including the actuators' masses), I_i is the rotational inertia of the i-th link and $q = [q_1 \quad q_2 \quad q_3 \quad q_4]^T$ contains the joint variables. Defining

$$\gamma := -\left(M_2 l_{c_2} l_1 \sin(q_2) + (M_3 + M_4) l_1 l_2 \sin(q_2)\right),$$

the matrix $C(q, \dot{q})$ can be written as follows:

$$C(q, \dot{q}) = \begin{bmatrix} \gamma\dot{q}_2 & \gamma(\dot{q}_1 + \dot{q}_2) & 0 & 0 \\ -\gamma\dot{q}_1 & 0 & 0 & 0 \\ 0 & 0 & 0 & 0 \\ 0 & 0 & 0 & 0 \end{bmatrix}.$$

The gravitational vector is

$$G(q) = \begin{bmatrix} 0 & 0 & -g(M_3 + M_4) & 0 \end{bmatrix}^T$$

where g is the gravitational acceleration.

In Tab. 3 we report the same parameters used in [10] for our simulations. These parameters have been previously taken from [9]. In Tab. 3, m_i denotes the saturation level of the i-th actuator.

Table 3. Parameters of the SCARA robot.

Link	l_i [m]	l_{c_i} [m]	M_i [kg]	I_i [kgm²]	m_i
1	0.6	0.3	12	0.36	55 Nm
2	0.4	0.2	6	0.08	60 Nm
3	1	$\frac{q_3}{2}$	3	0.08	70 N
4	0	0	1	0.08	25 Nm

Unconstrained controller design The unconstrained controller is a "computed torque" controller of the type (2), whose equations are not reported here for the sake of brevity with the following selection for the proportional, integral and derivative gains:

$$K_d = \mathrm{diag}(121.5, 30, 150, 150)$$
$$K_p = \mathrm{diag}(17.79, 8.25, 24.75, 20.13)$$
$$K_i = \mathrm{diag}(7.5, 10, 1, 0.5).$$

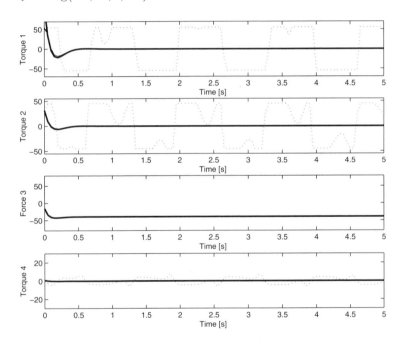

Fig. 8. SCARA robot: input responses to the reference (23) of the following closed-loop systems: unconstrained (bold solid), saturated (dotted), anti-windup from [10] (dashed) and new anti-windup law (thin solid).

Simulation results We report the simulations using two different anti-windup constructions. The first one is the original construction of [10], where the control law

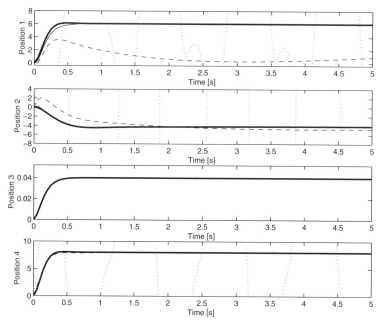

Fig. 9. SCARA robot: output responses to the reference (23) of the following closed-loop systems: unconstrained (bold solid), saturated (dotted), anti-windup from [10] (dashed) and new anti-windup law (thin solid).

(7) is used with the selection

$$K_g = \mathrm{diag}(0.9, 0.9, 0.4, 0.9)$$
$$K_0 = \mathrm{diag}(7.5, 4.5, 3.5, 2).$$

The second simulation corresponds to the new construction (10), (15) with the following selection for the parameters, which have been selected following similar procedures to those adopted in the previous two examples:

$$K_g = \mathrm{diag}(0.9, 0.9, 0.4, 0.9)$$
$$K_0 = \mathrm{diag}(60, 40, 30, 20)$$
$$K_q = \mathrm{diag}(280, 70, 70, 70).$$

In all the simulations both the plant and the controller states are initialized at zero.

We first reproduce the same simulation reported in [10], where the reference signal has been selected as

$$r = (6, -4, 4, 4) \ [\mathrm{deg, deg, cm, deg}]. \tag{23}$$

and both the plant and the controller states have been initialized at zero.

The corresponding responses are reported in Figs. 8 and 9.

Note that the new anti-windup law leads to extremely improved performance as compared to the previous law. The corresponding output response is almost coincident with the unconstrained trajectory thus providing almost full recovery of the original linear response. The unpleasant undershoot characterizing the previous anti-windup response from [10] has been completely eliminated and the unconstrained response recovery time almost reduced to zero (the response from [10] requires approximately 25 seconds to recover the unconstrained response on the first joint). Note also that for this simulation the saturated response leads to persistent oscillations (this was already observed in [10]).

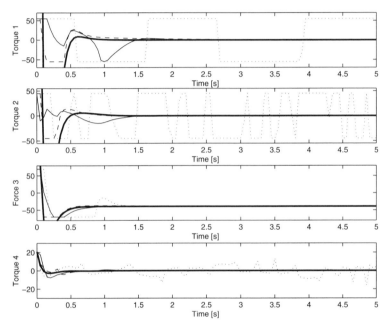

Fig. 10. Input responses to the reference (24) of the following closed-loop systems: unconstrained (bold solid), saturated (dotted), anti-windup from [10] (dashed) and new anti-windup law (thin solid).

Next, we report on a different experiment which is aimed at testing the reliability of the anti-windup law when the external reference corresponds to the following unreasonably high level:

$$r = (150, -100, 1, 200) \text{ [deg,deg,m,deg]}. \tag{24}$$

Note that in standard industrial manipulator controllers this set-point regulation task would be accomplished by generating a smoothened reference via cubic interpolation and verifying that the response does not exceed the saturation limits. However, we want to emphasize here that the system with anti-windup compensation does not require this particular action to take place and automatically exploits the full actuators

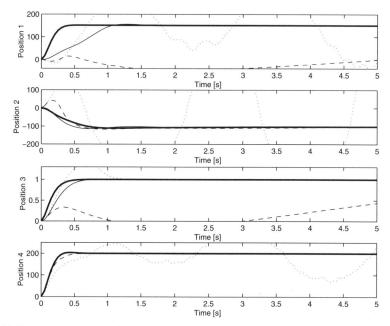

Fig. 11. Output responses to the reference (24) of the following closed-loop systems: unconstrained (bold solid), saturated (dotted), anti-windup from [10] (dashed) and new anti-windup law (thin solid).

power to guarantee a fast and graceful convergence to the desired set-point when driven by a simple step reference, regardless of its size. The resulting trajectories are reported in Figs. 10 and 11. In this case, as predictable, the saturated response (dotted) oscillates in an unreasonable way. However, also the anti-windup technique from [10] (dashed) provides poor performance, where the first three joints exhibit unacceptable undershoots and are associated with extremely slow transients. The new strategy (thin solid), instead, provides a response that almost coincides with the unconstrained one in the last three joints, while it is associated with a very fast transient on the first joint, requiring approximately 1.5 seconds to settle on the desired steady state. It is important to emphasize that different transients on each joint could be imposed by suitably adjusting the diagonal entries of the matrices K_q and K_0.

6   Conclusion

In this chapter we have proposed extensions of the anti-windup algorithm of [10], which lead to radical performance improvements of the compensated closed-loop behavior. Among other things, one advantage of the strategy here proposed is that the transient response of the anti-windup closed-loop system can be tuned by acting on simple decoupled proportional and derivative gains. The performance of the closed-loop system has been tested and verified by simulation on several examples. On the other hand, one disadvantage of this scheme is that it requires a model of the robot manipulator which must be inserted in the controller dynamics and may require a significant amount of computational effort.

Future research may include alternative selections for the nonlinear compensation law within the family of compensators which are proven here to stabilize the closed loop, as well as a formal proof which shows the ability of the anti-windup law to recover trajectory tracking properties of the controller under suitable assumptions on the trajectory to be tracked.

Acknowledgement

This work has been co-funded by AFOSR under grant F49620-03-1-0203 and NSF under grant ECS-0324679.

References

1. D. Angeli and E. Mosca, "Command governors for constrained nonlinear systems," *IEEE Trans. on Automatic Control*, vol. 44, pp. 816–820, 1999.
2. A. Bemporad, "Reference governor for constrained nonlinear systems," *IEEE Trans. on Automatic Control*, vol. 43, pp. 415–419, 1998.
3. C. Edwards and I. Postlethwaite, "Anti-windup and bumpless-transfer schemes," *Automatica*, vol. 34, pp. 199–210, 1998.
4. R. Hanus, "Antiwindup and bumpless transfer: A survey," *Proc. of 12th IMACS World Congress*, vol. 2, pp. 59–65, 1988.
5. Q. Hu and G.P. Rangaiah, "Anti-windup schemes for uncertain nonlinear systems," *IEE Proc. of Control Theory and Applications*, vol. 147, pp. 321–329, 2000.
6. N. Kapoor and P. Daoutidis, "An observer-based anti-windup scheme for non-linear systems with input constraints," *Int. J. of Control*, vol. 72, pp. 18–29, 1999.
7. T.A. Kendi and F.J. Doyle, "An anti-windup scheme for multivariable nonlinear systems," *J. of Process Control*, vol. 7, pp. 329–343, 1997.
8. M.V. Kothare, P.J. Campo, M. Morari, and N. Nett, "A unified framework for the study of anti-windup designs," *Automatica*, vol. 30, pp. 1869–1883, 1994.
9. G. Mester, "Adaptive force and position control of rigid-link flexible-joint SCARA robots," *Proc. of 20th IEEE Industrial Electronics Conference*, pp. 1639–1644, 1994.
10. F. Morabito, A.R. Teel, and L. Zaccarian, "Anti-windup design for Euler-Lagrange systems," *Proc. of 2002 IEEE Int. Conf. on Robotics and Automation*, pp. 3442–3447, 2002.
11. A.R. Teel and N. Kapoor, "Uniting local and global controllers," *Proc. of 4th European Control Conf.*, 1997.
12. S. Valluri and M. Soroush, "Input constraint handling and windup compensation in nonlinear control," *Proc. of 1997 American Control Conf.*, 1997.

Model-Based Friction Compensation

Gianni Ferretti, Gianantonio Magnani, and Paolo Rocco

Dipartimento di Elettronica e Informazione
Politecnico di Milano
Piazza Leonardo Da Vinci 32, 20133 Milano, Italy
<*ferretti,magnani,rocco*>*@elet.polimi.it*
http://www.elet.polimi.it/upload/ferretti/metromod

Abstract. Compensation of nonlinear friction terms is a most challenging application of high resolution encoders, which are nowadays getting available for common industrial motion control and robotic applications. In fact, use of a high resolution sensor allows a neat analysis of the dynamic behavior of friction forces in the presliding regime, and especially of hysteresis loops. Starting from a recently proposed friction model, defining more accurately the presliding regime, a research is presented in this chapter, aimed at devising identification and compensation procedures for friction.

1 Introduction

Friction appears in all mechanical systems and is a major source of control performance degradation. Its worst effects are observed in the form of static errors, limit cycles, stick-slip motions, as well as quadrant glitches [1,2,6].

Some of these effects have been eliminated by superimposing dither signals on the commands generated by the controller, or by closing an acceleration feedback. These techniques avoid the need of deriving a model of friction, generated by several complicated physical mechanisms. On the contrary, the topic of this work is model-based friction compensation, whose block diagram scheme is reported in Fig. 1.

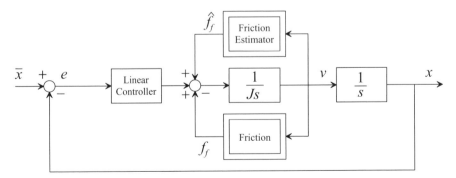

Fig. 1. Block diagram of the model-based friction compensation scheme.

B. Siciliano et al. (Eds.): Advances in Control of Articulated and Mobile Robots, STAR 10, pp. 87–100, 2004.

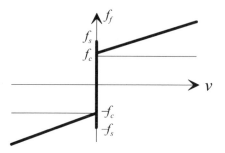

Fig. 2. Classical model.

The main problems with friction take place at velocity reversal, where the classical friction model considers a discontinuity in the friction force(torque)/velocity characteristics, shown in Fig. 2 (for the sake of simplicity a symmetric characteristics is assumed throughout the chapter, even if generally friction is different in the two directions of motion). This characteristics however defines uniquely the friction force only for $v \neq 0$. In this case it is

$$f_f = \sigma_2 v + f_c \text{sgn}(v) \tag{1}$$

where σ_2 is the viscous friction coefficient and f_c is the Coulomb friction.

When $v = 0$ the characteristics just establishes that $f_f < f_s$, with f_s being the *stiction* force. To precisely determine the friction force an additional variable has therefore to be considered, the net active force f_a, namely the algebraic sum of the forces acting on the mobile body (assuming for simplicity that only one body is mobile, the others being fixed) apart from friction. Thus, in rest conditions

$$f_f = f_a \tag{2}$$

as shown in Fig. 3, where it is also pointed out that Eq. (2) holds for $|f_a| \leq f_s$. Note that Eq. (2) simply states $dv/dt = 0$.

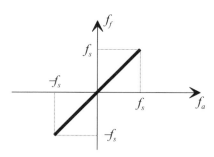

Fig. 3. Friction force at rest.

The correct implementation of the classical friction model is therefore much more complicated than (1) and it is performed in [5] through a finite state machine, distinguishing three states: motion, stiction and incipient motion. The model is aimed at detecting precisely the instant of motion stop, overcoming the numerical problems related to the well-known Karnopp model [15], which assigns a null value to the $\text{sgn}(v)$ function over a suitable short interval around zero.

More accurate friction models have been recently proposed in the literature, introducing two different motion regimes, *sliding* and *presliding*, and overcoming the discontinuity of the classical model by introducing a relation between friction force and relative displacement in the presliding regime. These models account for the *microsliding displacements* observed at motion start or reversal with high resolution measurement systems. A comprehensive survey of friction models is reported in [20], where an evolution from static to dynamic models is pointed out.

The first dynamic friction model has been proposed in [9], starting from the stress-strain curve of classical solid mechanics, modeled by a differential equation:

$$\frac{df_f}{dx} = \sigma_0 \left(1 - \frac{f_f}{f_c}\text{sgn}(v)\right)^{\alpha}, \tag{3}$$

where x is the relative displacement, σ_0 is the stiffness coefficient, and α is a parameter that determines the shape of the stress-strain curve. The behavior of the so-called Dahl model can be visualized as in Fig. 4. The contact is modelled as

Fig. 4. Dahl model.

occurring at some *junctions*, formed under the action of a normal load. For small (micro) relative displacements between the two contacting surfaces these junctions behave as linear springs, generating the friction force. When the friction force reaches a maximum the spring breaks, and the sliding motion starts. Junctions form instantaneously when the relative motion stops. The maximum value of the friction force and the maximum displacement are also called respectively *breakaway* force and displacement. Typical values of the breakaway displacements are in the order of $2 \div 5$ μm for steel junctions [1]. A time domain model can be easily derived from (3) as

$$\frac{df_f}{dt} = \sigma_0 \left(1 - \frac{f_f}{f_c}\text{sgn}(v)\right)^{\alpha} v,$$

which, for $\alpha = 1$ reduces to

$$\frac{dz}{dt} = v - |v|\frac{\sigma_0 z}{f_c} \tag{4}$$

$$f_f = \sigma_0 z , \tag{5}$$

where the state variable z has been introduced. This state variable can be also interpreted as the average deflection of elastic *bristles*, deflecting under the action of a tangential force [13]. When a maximum deflection $z_{ss} = f_c/\sigma_0$ is reached, corresponding to a maximum friction force $f_f = f_c$, the sliding motion starts.

The Dahl model describes properly the presliding regime, which macroscopically appears as an abrupt start and stop of the relative motion, but does not account for the behavior of friction during sliding. To this aim, a modification of the Dahl model was proposed in [4] which, however, does not account explicitly for the relative velocity. This appears to be essential, also considering the effect of lubrication [1].

Grease or oil lubrication has the main purpose of creating a fluid film between the two contacting surfaces, avoiding solid-to-solid contact. Generally hydrodynamic lubrication is performed, thus the lubricant is pushed into the contact zone by the relative velocity. There are four regimes of lubrication (Fig. 5):

I. Static friction The static friction regime is well described by the Dahl model.

II. Boundary lubrication In the boundary lubrication regime the relative velocity is not adequate to build a fluid film between the contacting surfaces. As such, friction is generally higher than for fluid lubrication (regimes III and IV).

III. Partial fluid lubrication In the partial fluid lubrication regime some lubricant is drawn into the contact zone and some is expelled by the load pressure; the greater viscosity or motion velocity, the thicker the fluid film. In this regime however the film is not sufficiently thick and some solid-to-solid contact still holds.

IV. Full fluid lubrication When the film is sufficiently thick, the separation of the surfaces is complete and the load is fully supported by the fluid. In this regime the viscosity of the lubricant dominates and friction increases with velocity.

Particularly important, for its influence on the rising of stick-slip motions, is the so-called Stribeck effect [22], namely the regime of decreasing friction with increasing velocity at low velocity (negative viscous friction, between regime II and III in Fig. 5). The dependence of the friction force from velocity, Stribeck effect included, can be parameterized as follows:

$$f_f = \text{sgn}(v)h(v) \tag{6}$$

$$h(v) = f_c + (f_s - f_c)\exp\left[-(|v|/v_s)^\delta\right] , \tag{7}$$

where v_s is the Stribeck velocity and δ is a suitable parameter.

There are also two important *temporal* phenomena [21], not considered in this work: a relation between the time spent in the stuck condition, or *dwell time*, and the level of static friction, and a delay between a change in velocity and the corresponding

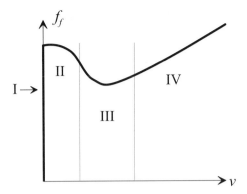

Fig. 5. Lubrication regimes.

change in friction, or *frictional lag*. In [16] the following empirical model relating static friction and dwell time has been proposed:

$$f_s(t) = f_{s,\infty} - (f_{s,\infty} - f_{c,k}) \exp\left(-\gamma t^m\right)$$

where $f_{s,\infty}$ is the asymptotic static friction, $f_{c,k}$ is the Coulomb friction at the moment of arrival in the stuck condition and γ, m are empirical parameters. The frictional lag seems to be related to the time required to modify the lubricant film thickness. Experimental data suggest a simple time delay as a model of this process [14].

The first integrated model, accounting for both the sliding and presliding regime, was proposed in [8], which combined the Dahl model (4,5) (with $\alpha = 1$) with (6,7) (with $\delta = 2$) into the well known LuGre model:

$$\frac{dz}{dt} = v - |v|\frac{\sigma_0 z}{h(v)} \tag{8}$$

$$h(v) = f_c + (f_s - f_c) \exp\left[-(v/v_s)^2\right] \tag{9}$$

$$f_f = \sigma_0 z + \sigma_1 \frac{dz}{dt} + \sigma_2 v . \tag{10}$$

They also introduced a micro-viscous friction term, proportional to the time derivative of the state variable z through the coefficient σ_1. Conditions for passivity of the model have been discussed in [20] and in [3]. The LuGre model has been also extended to model point contact in grasping tasks [12]. The model is local to the point of contact and is applicable to an arbitrary number of contacts among fingers and grasped object. The LuGre model is very elegant and easy to implement and lends itself to use in adaptive friction compensation schemes [7].

Recently, however, the model has been shown to exhibit a nonphysical drift phenomenon, which originates from modelling presliding as a combination of elastic and plastic displacements [10]. Moreover, on the ground of experimental observations, the LuGre model has been subject to several criticisms in [23], mainly addressing the hysteresis behavior in presliding. In particular, it is remarked that the LuGre model does not account for nonlocal memory and cannot accomodate arbitrary displacement-force transition curves. A hysteresis behavior with nonlocal memory is defined as an input-output relationship where the output not only depends on the input and the output at some time instant in the past, but also on past extremum values of the input or output as well [19]. A a new friction model, the so-called Leuven model, has been proposed as

$$\frac{dz}{dt} = v \left(1 - \operatorname{sgn} \left(\frac{f_h(z)}{s(v)} \right) \left| \frac{f_h(z)}{s(v)} \right|^n \right) \tag{11}$$

$$s(v) = \operatorname{sgn}(v) h(v) \tag{12}$$

$$f_f = f_h(z) + \sigma_1 \frac{dz}{dt} + \sigma_2 v. \tag{13}$$

where n is a coefficient used to shape the transition curves and f_h is the hysteresis force, i.e. the part of the friction force exhibiting hysteresis behavior with state variable z as input. It consists of two parts

$$f_h(z) = f_b + f_d(z),$$

where $f_d(z)$ is a point-symmetrical strictly increasing function of z, to be determined experimentally, while f_b is the value of $f_h(z)$ at the beginning of a transition curve.

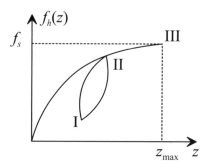

Fig. 6. Hysteresis loops.

The implementation of $f_h(z)$ requires two memory stacks, one for the minima of f_h (stack min) and one for the maxima (stack max), which grow and shrink according to the following rules (Fig. 6):

I. Velocity reversal At velocity reversal a new transition curve is started and a new extreme value of f_h has to be added to one of the stacks.
  1. The former value of $f_h(z)$ is placed on the stack max/min in the case of going from positive/negative to negative/positive velocity and becomes the new value of f_b.
  2. The state variable z is reset to 0.

II. Closing an internal loop The closed hysteresis loop is removed from the hysteresis memory (wiping out).

1. The elements on the stacks associated with the internal loop are removed.
2. The new value of f_b is the top value on the stack min/max for positive/negative velocity.
3. The value of z is recalculated such that a transition curve starts at the new value of f_b while maintaining the continuity of f_h.

III. Transition from presliding to sliding The hysteresis model is reset for strictly positive/negative velocities, when the hysteresis force reaches a maximum or a minimum ($f_h \approx \pm f_s$).

1. The stacks are cleared out and their first elements are set to $\pm f_s$.
2. f_b is set to $-f_s$ (f_s) for positive (negative) velocities.
3. The value of z is recalculated so as to maintain the continuity of f_h.

The Leuven model allows a very accurate modeling of friction, particularly in the presliding regime, but the stacks mechanism is quite cumbersome to be implemented in real time and may result in overflow. In fact, several velocity reversals may occur without closing of inner loops, causing the growth of the stacks, whose size must be chosen in advance.

The stack overflow problem has been overcome in a further refinement of the Leuven model [18], by modeling the hysteresis force through the Maxwell slip model. The model is defined by N massless elastoslide elements in parallel (Fig. 7(a)).

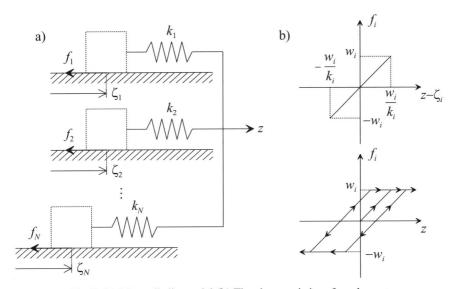

Fig. 7. (a) Maxwell slip model (b) The characteristics of an element.

Each element i has one common input z and one output f_i and is characterized by a maximum force w_i, a linear spring constant k_i and a state variable ζ_i, describing the position of element i (Fig. 7(b)). The behavior of each element is described as follows:

$$\text{if } |z - \zeta_i| < \frac{w_i}{k_i} \text{ then } \begin{cases} f_i = k_i(z - \zeta_i) \\ \zeta_i = \text{const} \end{cases}$$

$$\text{else } \begin{cases} f_i = \text{sgn}(z - \zeta_i)w_i \\ \zeta_i = z - \text{sgn}(z - \zeta_i)\frac{w_i}{k_i} \end{cases} .$$

The hysteresis force is equal to the sum of hysteresis forces f_i of each element:

$$f_h = \sum_{i=1}^{N} f_i .$$

The model works as follows. When the relative motion stops ($v = 0$) all elements are sticking and the total stiffness will be the sum of the stiffnesses of all elements. When the force f_i reaches the saturation level w_i, the i-th element starts to slip and the total stiffness decreases with the stiffness of the spring element i.

The last version of the Leuven model has been considered in this chapter for implementing model-based friction compensation. The model is first identified and validated in Section 2 on an experimental setup, while some experimental results, obtained with a feedforward compensation, are discussed in Section 3. Section 4 finally draws some conclusion and perspectives for future research.

2   Identification and Validation of the Model

The experimental test bed adopted is shown in Fig. 8. It is made up by a brushless motor (Control Techniques UNIMOTOR), a harmonic drive gearbox (model HFUC size 25), with a gear ratio $n = 100$ and a fully digital drive and actuation system (Control Techniques). The motor angle is sampled at a frequency of 4 KHz, with a resolution of 22 bits/round, namely more than 4 million pulses per revolution, 1.5 μrad or 8.6×10^{-5} deg. The velocity is estimated by numerical derivation. The motor torque u is also not measured directly but is estimated from the current setpoint $\bar{I}$ as $u = K_t \bar{I}$.

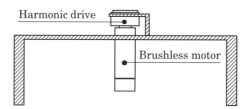

Fig. 8. Experimental test bed.

The friction parameters relative to the sliding regime for the test bed adopted were already identified in [11]. In this respect it must be pointed out that while a symmetric friction characteristics is assumed in (7), different values were identified for positive and negative rotations. This fact is neglected in this chapter, where the main focus is on the identification and compensation of friction at motion reversal, i.e. of static friction, and a mean amplitude value for the sliding friction parameters is assumed.

The identification of $f_h(z)$ has been performed as in [23], thus applying a slow current ramp (in presliding regime) and considering the following relation:

$$\frac{dz}{dt} = v \left(1 - \left(\frac{f_h}{f_s} \right)^n \right) , \tag{14}$$

in order to calculate z, with $n = 7$. However, differently from [23], the dynamical effects were taken into account for the calculation of f_h. The friction torque was in fact computed as

$$f_h = K_t \bar{I} - J \dot{v} . \tag{15}$$

where J is the motor inertia.

As far as the identification of f_h is concerned, it must be pointed out that choosing the values of w_i and k_i in order to approximate the real hysteresis is a nonlinear problem. If the maximum deflection of each element $\Delta_i = w_i/k_i$ is pre-assigned, in place of k_i, the identification model can be rewritten as a nonlinear equation which is linear in the unknown parameters w_i:

$$f_h(k) = \sum_{i=1}^{N} w_i \Phi_i(z(k), \xi_i(k), \Delta_i(k)) ,$$

and can be therefore identified using a least squares method [17].

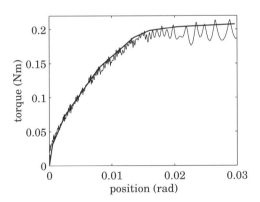

Fig. 9. Friction characteristics.

Figure 9 shows the characteristics $f_h(z)$, computed from (14) and (15) (thin line), and the piecewise linear function implementing the Maxwell slip model (thick line) in this particular experiment. In fact, it is possible to figure out that the springs contributing to the overall characteristics $f_h(z)$ break one at a time, after an elongation Δ_i, exerting a constant force w_i after breakdown. Accordingly, a piecewise linear characteristics is obtained, whose slope ranges from a maximum value at motion inversion, given by $\sum_{i=1}^{N} k_i$, to a minimum value k_N.

In order to determine the values of the spring constants k_i, the estimated $f_h(z)$ has been first averaged, so as to eliminate the fluctuations due to the acceleration term in (15). Then, a number of N maximum elongations Δ_i has been suitably chosen and the values of k_i have been calculated:

$$k_N = \frac{\bar{f}_h(\Delta_N) - \bar{f}_h(\Delta_{N-1})}{\Delta_N - \Delta_{N-1}}$$

$$k_{N-i} = -\sum_{j=N-i+1}^{N} k_j + \frac{\bar{f}_h(\Delta_{N-i}) - \bar{f}_h(\Delta_{N-i-1})}{\Delta_{N-i} - \Delta_{N-i-1}} \qquad i = 1, \ldots, N-1 .$$

With $\Delta_0 = 0$, $f_h(0) = 0$, $f_h(\Delta_N) = f_s$. On-line, recursive identification of the model has been proposed in [17].

Some experiments were afterwards performed in order to assess the validity of the model, in particular in replicating the hysteresis cycles. Some results are shown in Fig. 10.

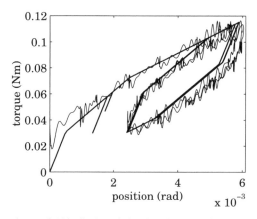

Fig. 10. Experimental (thin line) and simulated hysteresis cycles (thick line).

3   Friction Compensation: Experimental Results

A feedforward compensation has been applied (Fig. 11), considering three sinusoidal velocity profiles $\bar{v}(t) = A_0 + A_1 \sin(\omega t)$.

In a first experiment the following values were chosen: $A_0 = 0$ $A_1 = 180$ rad/s, $\omega = 3\pi/5$ rad/s, $J = 1.9 \times 10^{-4}$ Kgm^2 and the velocities obtained with and without

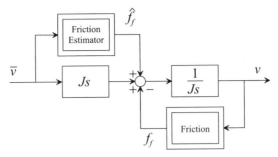

Fig. 11. Block diagram of the feedforward friction compensation scheme.

compensation, together with the nominal velocity profile (dotted line), are reported in Fig. 12. Note that the friction model did not take into account the sliding regime.

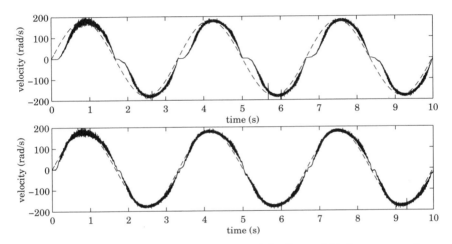

Fig. 12. Velocities without (top) and with (bottom) feedforward friction compensation.

In a second experiment a slower sinusoid was considered: $\omega = 3\pi/10$ rad/s, in order to emphasize the occurrence of stiction. In this case the effect of the compensation is even more evident (Fig. 13).

It must be pointed out that in the above experiments the velocity changes sign after stiction, as such $f_h(z)$ increases (decreases) monotonically from $-f_s$ $(+f_s)$ to $+f_s$ $(-f_s)$ and no hysteresis cycle occurs. In order to evaluate the performance of the compensation even when motion stops and restarts in the same direction an experiment was performed with $A_0 = A_1 = 108$ rad/s, $\omega = 3\pi/5$ rad/s. The effect of the compensation is again evident (Fig. 14): no stiction occurs and the actual velocity follows better the nominal velocity in case of compensation. Note however that the feedforward model does not exactly predict the actual instant of vanishing velocity, so that overcompensation occurs.

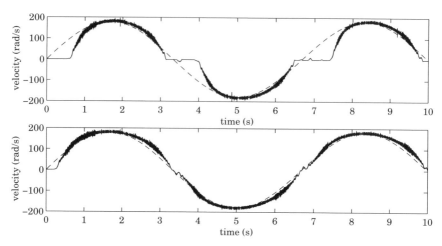

Fig. 13. Velocities without (top) and with (bottom) feedforward friction compensation.

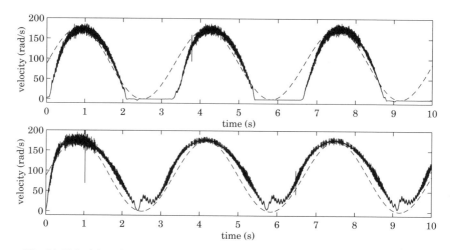

Fig. 14. Velocities without (top) and with (bottom) feedforward friction compensation.

4   Conclusion

In this chapter, a recently proposed model of friction (the Leuven/Maxwell slip model) is considered as the starting point for the investigation of friction compensation techniques. The model is particularly accurate in the presliding regime, where the worst effects of friction take place. In particular, the model correctly predicts the hysteresis cycles in the characteristics relating the friction force to the model state variable, representative of the microsliding displacements.

In order to appropriately apply the model, however, a high resolution position measurement is needed, here ensured by the adoption of a commercial encoder with a resolution of 22 bit per turn.

The Leuven/Maxwell slip model model has been identified and validated on an experimental test bed and some promising results have been obtained with a feedforward compensation of friction. A first implementation of a feedback compensation however failed because the computational burden entailed by the outlined model was incompatible with the short sampling time ($250\ \mu s$).

The next steps of research are the implementation of the feedback compensation, which is expected to require some modifications of the model, mainly in order to improve the computational efficiency, and the development of some adaptation techniques, in order to deal with the variations of friction with load, time and temperature.

References

1. B. Armstrong-Hélouvry, *Control of Machines with Friction*, Kluwer Academic Publishers, 1991.
2. B. Armstrong-Hélouvry, P. Dupont, and C. Canudas de Wit, "A survey of models, analysis tools and compensation methods for the control of machines with friction," *Automatica*, vol. 30, pp. 1083–1138, 1995.
3. N. Barabanov and R. Ortega, "Necessary and sufficient conditions for passivity of the LuGre friction model," *IEEE Trans. on Automatic Control*, vol. 45, pp. 830-832, 2000.
4. P.-A. Bliman and M. Sorine, "Easy-to-use realistic dry friction models for automatic control," *Proc. of 3rd European Control Conf.*, pp.'3788–3794, 1995.
5. A. Bonsignore, G. Ferretti, and G. Magnani, "Analytical formulation of the classical friction model for motion analysis and simulation," *Mathematical and Computer Modelling of Dynamical Systems*, vol. 5, pp. 43–54, 1999.
6. A. Bonsignore, G. Ferretti, and G. Magnani, "Coulomb friction limit cycles in elastic positioning systems," *ASME J. of Dynamic Systems, Measurement, and Control*, vol. 121, pp. 298–301, 1999.
7. C. Canudas de Wit and P. Lischinsky, "Adaptive friction compensation with partially known dynamic friction model," *Int. J. of Adaptive Control and Signal Processing*, vol. 11, pp. 65–80, 1997.
8. C. Canudas de Wit, H. Olsson, K.J. Åström, and P. Lischinsky, "A new model for control of systems with friction," *IEEE Trans. on Automatic Control*, vol. 40, pp. 419–425, 1995.
9. P.R. Dahl, *A Solid Friction Model*, Tech. Rep. TOR-158(3107-18), Aerospace Corporation, El Segundo, CA, 1968.
10. P. Dupont, V. Hayward, B. Armstrong, and F. Altpeter, "Single state elastoplastic friction models," *IEEE Trans. on Automatic Control*, vol. 47, pp. 787-792, 2002.
11. G. Ferretti, G. Magnani, G. Martucci, P. Rocco, and V. Stampacchia, "Friction model validation in sliding and presliding regimes with high resolution encoders," in B. Siciliano and P. Dario (Eds.), *Experimental Robotics VIII*, pp. 328–337, Springer Verlag, 2003.
12. G. Ferretti, G. Magnani, and P. Rocco, "Modular dynamic modeling and simulation of grasping," *Proc. of 1999 IEEE/ASME Int. Conf. on Advanced Intelligent Mechatronics*, pp. 428–433, 1999.
13. D.A. Haessig and B. Friedland, "On the modeling of simulation of friction," *ASME J. of Dynamic Systems, Measurement, and Control*, vol. 113, pp. 354–362, 1991.
14. D.P. Hess and A. Soom, "Friction at a lubricated line contact operating at oscillating sliding velocities," *J. of Tribology*, vol. 112, pp. 147–152, 1990.

15. D. Karnopp, "Computer simulation of stick-slip friction in mechanical dynamic systems," *ASME J. of Dynamic Systems, Measurement, and Control*, vol. 107, pp. 100–103, 1985.
16. S. Kato, N. Sato, and T. Matsubay, "Some considerations of characteristics of static friction of machine tool slideway," *J. of Lubrication Technology*, vol. 94, pp. 234–247, 1972.
17. V. Lampaert and J. Swevers, "On-line identification of hysteresis functions with nonlocal memory ," *Proc. of 2001 IEEE/ASME Int. Conf. on Advanced Intelligent Mechatronics*, pp. 833–837, 2001.
18. V. Lampaert, J. Swevers, and F. Al-Bender, "Modification of the Leuven integrated friction model structure," *IEEE Trans. on Automatic Control*, vol. 47, pp. 683–687, 2002.
19. I.D. Mayergoyz, *Mathematical Models of Hysteresis*, Springer Verlag, 1991.
20. H. Olsson, K.J. Åström, C. Canudas de Wit, M. Gäfvert, and P. Lischinsky, "Friction models and friction compensation," *European J. of Control*, vol. 4, pp. 176-195, 1998.
21. E. Rabinowicz, "The intrinsic variables affecting the stick-slip process," *Proc. of Physical Society of London*, vol. 71, pp. 668–675, 1958.
22. R. Stribeck, "Die wesentlichen Eigenschaften der Gleit- und Rollenlager," *Zeitschrift des Vereines Deutscher Ingenieure*, vol. 46, pp. 1342-1348, 1432-1437, 1902.
23. J. Swevers, V. Lampaert, F. Al-Bender, and T. Prajogo, "An integrated friction model structure with improved presliding behavior for accurate friction compensation," *IEEE Trans. on Automatic Control*, vol. 45, pp. 675–686, 2000.

Architectures for Rapid Prototyping of Model-Based Robot Controllers

Basilio Bona, Marina Indri, and Nicola Smaldone

Dipartimento di Automatica e Informatica
Politecnico di Torino
Corso Duca degli Abruzzi 24, 10129 Torino, Italy
<*basilio.bona,marina.indri,nicola.smaldone*>*@polito.it*
http://www.ladispe.polito.it/robotica/labrob

Abstract. Rapid Prototyping (RP) in control design can be defined as a computer-assisted process aimed at recursively validating dynamic models of complex plants and mechatronic systems and/or designing and testing digital control algorithms for real-time applications. Rapid prototyping of digital control algorithms requires integrated hardware/software architectures, allowing fast and systematic interactions between the algorithmic design phase and the experimental testing. The design phase is performed with the support of a computer-aided control design environment, where simulations are performed on accurate models of the specific equipment under investigation; after that, a rapid transfer of the algorithm on the target hardware is necessary to validate it experimentally. It is therefore necessary to have a complete prototyping environment, where different controller blocks are readily available, with structure and parameters easily modifiable to be tested on the simulated plant and downloaded on the target hardware platform for real-time validation. The present chapter introduces the state of the art on RP, critically surveys and discusses general issues related to both HW and SW aspects that are at the basis of RP; furthermore it describes in some details the solution implemented by the authors at the Experimental Robotics Laboratory of Politecnico di Torino. A test case, devoted to the problem of modelling and compensation of nonlinear friction in rotating robot arms is presented. Finally, a critical appraisal of the proposed solution, in the light of the gained experience, is discussed and future developments are pointed out.

1 Introduction

Prototyping can be defined as: "A type of development in which emphasis is placed on developing prototypes early in the development process to permit early feedback and analysis in support of the development process" [4]. The implementation of a prototype starts from an idea which is then developed in a project phase, where several alternative solutions are considered to achieve the desired functionalities and specifications. Design relies on technical competence and objectives; several tools can help the designer to practice that competence and to define the objectives in details. Using a Personal Computer (PC) in the prototyping phase as a replacement of traditional technical tools is a common practice today. One of the most important features of the PC is the possibility of virtualizing the objects and the procedures to build them. For example, in architectural design, the computer graphics allows visualizing the whole inhabited environment in some details and verifying the design

B. Siciliano et al. (Eds.): Advances in Control of Articulated and Mobile Robots, STAR 10, pp. 101–123, 2004.
© Springer-Verlag Berlin Heidelberg 2004

hypothesis formulated by the architect; mechanical engineers can try to combine some graphical objects representing mechanical parts, starting from the drawings of such parts, to test their functionalities.

A major interest in prototyping derives from the possibility of knowing the influence of design solutions before the final production phase. In manufacturing, where small technological objects are often produced in large quantities, prototyping allows building the trial version in order to verify a subset of functionalities, before the cost of possible design errors grows up due to the large number of manufactured parts. In such a situation the PC can be useful as it automates the large number of procedures involved in the construction of the prototype.

In more general terms, the prototyping process makes easier the application of specific methodologies from different technological fields embedded into a real or virtual instance of a product.

In the last few years these aspects are becoming one of the major issues in control design for advanced mechatronic equipments and robotics [2,7,10].

In the field of industrial robotics there are several kinds of prototyping processes; a manipulator embodies different technologies and competences: mechanical, electronic and electrical issues merge with automatic control and computer science competencies for a satisfactory design of the whole machine.

In the present work we discuss prototyping issues and architectures for control and supervision of industrial robots; the aim of prototyping is often the implementation of new control algorithms or architecture allowing better performances at lower costs in well defined operating conditions.

An objective only partially reached today is the so-called **rapid prototyping**, i.e. a methodology which allows going in a short time and with limited costs from the general idea to the realizable solution. After the prototype design is tested on the real equipment, one must be able to repeat cyclically the same procedure with only a marginal additional effort.

The prototyping process consists of a set of phases, often technologically very different; this fact complicates in a remarkable way the transmission of information, especially when formalisms and techniques used before the PC advent are involved. So, rapid prototyping must be based on a friendly and homogenous development environment, which should allow the designer to concentrate on conceptual problems freeing him/her from the tedious practical aspects involved in the progression from the idea to the prototype.

The PC plays a major role hosting the interactive environment allowing to develop many of the rapid prototyping process phases, such as for example:

- to model the controlled electromechanical parts,
- to design the control laws and the machine supervision software,
- to simulate the effects of the control algorithms,
- to automate the transition from the design formalism to the implementation and adaptation to the machine execution,
- to manage the interaction between designer and test machine.

The last point introduces a problem that is common to all the environments where automatic controlled evolution of physical phenomena is needed, i.e. real-time requirements.

Industrial robots are supplied with a controller cabinet containing hardware and software systems for control and supervision. Due to industrial secrecy, safety requirements or, sometimes, technological backwardness, these systems are closed to modification by the customers. On the other hand a controller presents many critical aspects due to the simultaneous presence of components with contrasting real-time requirements.

The user of a prototyping system should have at least the possibility to interface the original control environment, and in many cases partially or totally substitute it; therefore it is necessary to pay attention to the real-time issues in order to avoid interferences with the native architecture, especially when it is necessary to replace important functionalities.

In the following section some concepts related to the real-time interaction between PC and controlled mechatronic equipments will be introduced; basic definition will be briefly presented, and methodologies will be described. Particular attention will be paid to the real-time requirements of rapid prototyping systems.

2   Rapid Prototyping

For each specific "product" the prototyping process requires a test platform, where it is possible to investigate the characteristics and the potentiality of several alternative solutions, before arriving to the final product release. In the field of control systems for industrial robots, the product usually consists in control algorithms for robotic axes or software for machine supervision and man-machine interface; at the same time this platform allows dealing in an efficient manner with the plant modelling too, using simple simulation and test procedures.

The electronics of a control system ready to be commercialized is the result of optimization in workspace, performance, reliability and costs. It is advisable to test the functionalities of the design ideas using standard and re-usable components; on a single prototype costs are often greater than those of the final product, but the possibility of searching a solution without worrying about non-functional constraints (power consumptions, space, reliability, etc.) and the simplification of more complex problems should be considered. Furthermore, the product will be often sold in large quantities, making the prototype costs negligible. In other cases the costs may be disregarded because the designer is interested only in a limited number of performances with respect to those of the actual product.

The test platform usually consists of a software environment representing the plant and the control components according to some conceptual metaphor, often a graphical one. One of the main characteristics in this environment is the possibility to simulate the functionalities of the system both on time basis and on logical basis.

Often the prototyping software is coupled with the real plant using configurable electronic boards. The advantages of this methodology result from two fundamental factors: a) the extreme configurability of the software environment, which allows modifying all the design parameters and to foresee the possible consequences; b) the adaptability of the prototyping electronics composed by standard and modular components.

The possibility of simulating the control algorithms avoids potentially dangerous situations for the equipments, which, on the other end, may be inaccessible for the tests. Indeed, the complexity of some plants requires a separation of the design into subsystems, where each one needs to be tested independently from the others. This procedure may be impossible for industrial manipulators with many degree of freedom due to the highly coupled nature of the kinematic chain.

In industrial manipulator prototyping the possibility of simply formalizing the mathematical models for kinematics and dynamics, which will be used in the simulation software, is one of the more interesting features; in this manner all the control algorithms can be tested and interfaced with these models. Models must be enough refined to describe all the phenomena judged critical for control; for example, the model can take into account joint frictions, disturbances and parameters uncertainty, but not address the elasticity issues, if these are not critical.

When the robot is accessible and it is possible to interface it with prototyping system, the test phase can be managed directly from the development and simulation environment. The prototyping system is sometimes called **Host** and it is independent from the constraints due to the plant interfacing, thanks to the presence of another computer, the **Target**, which supervises to the interaction with the manipulator. The Host interacts in asynchronous mode with the Target to set the test execution modalities and to monitor its progression; the Target receives commands and data from the Host and, through the interface electronics, controls the robot in real-time. If the plant is particularly articulated, then a multi-Target system, possibly with each Target synchronized with the others, can be necessary, whereas the Host can remain unique. In some particular cases, a single system can be used as both development and plant interaction environment.

The Host environment is often called a **CAD system** since it allows a "Computer Aided Design"; in particular, for automatic controls it is named **CACSD**, from "Computer Aided Control System Design". The Target environment can have one of the possible architectures suitable for real-time requirements; these architectures and some basic concepts are presented in the following section, on the basis of quantity and complexity of tasks concentrated on it.

2.1   Real-Time Systems

The key issue of real-time systems for automatic control is the proper interaction with physical phenomena representing real processes.

Interaction, performed by a computer program, takes place through signals, whose time history is characterised by a dominating time constant. Because of the limits on the available resources, it is necessary to select the relevant time constants, in order to update the knowledge of input signals, and to recompute the output commands to control the process as requested.

A real-time system uses software structures called **Tasks** to perform this kind of interaction while complying with the assumed time constraints.

These Tasks are often characterized by different time constants and have to be executed within the same time window by the same computer; for this reason the software needs to share the calculus resources between all these activities.

A Task with **hard real-time** requirements must complete its job strictly inside the time interval planned on the basis of the control criticalities, in order to avoid the total failure of the process. So it is necessary to be certain, using some procedures or exhaustive simulations, that the system will not infringe the time constraints of that Task [9].

A Task with **soft real-time** requirements, instead, lacks the claim of never infringe the time constraints, or warrants it in some statistical sense, e.g. in the "majority of cases": infringing the time constraints for the soft components is accepted as an unsubstantial degradation of the system functionalities.

Both these kinds of Tasks can coexist in a robot control system; for example, the closed loops of the control axis or the emergency procedures related to the limit switches activation are hard real-time Tasks, whereas signaling non-critical anomalies or refreshing the control workstation graphical user interface are considered soft real-time Tasks.

In this context an implicit hypothesis is assumed: all Tasks are arranged on a unique computer and the capacity of parallel execution of different connected Tasks is called **multitasking**. The hard real-time requirements fulfillment can be guaranteed avoiding to use hardware and software components which can bring to non-deterministic behaviors; or, according to more refined techniques, dynamically checking that each new Task will have the possibility, on the basis of the current workload, of completing its jobs and respecting its time constraints.

2.2 Architectures, Characteristics and Requirements of Robot Prototyping Systems

The hardware architecture of a simple **Commercial-Off-The-Shelf** (COTS) computer is shown in Fig. 1.

RAM and CPU are the fundamental resources, and various Tasks use them according to defined specifications and procedures, aimed to implement a correct multitasking for real-time requirements. To allow the Task interaction with the outer world and the physical phenomena under control, additional components, called Input/Output devices, are needed. These components are essentially electronic devices acting as bit converters to and from electrical external signals. A typical example of I/O devices for automatic control systems are DACs, Digital-to-Analog Converters, and ADCs, Analog-to-Digital Converters.

Communication between CPU, RAM and I/O devices is carried out through a system Bus, which becomes a shared resource and, as such, requires a sharing mechanism. It should be noted that each time resources are shared, the time constraints and the deterministic behavior dictated on the real-time system are threatened: if two

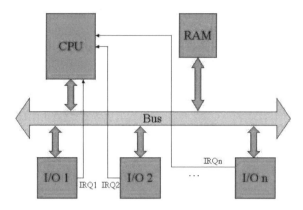

Fig. 1. A very simple elaboration system.

real-time Tasks need the same data bus, the first which obtain it can delay or prevent the other Task to complete its job in time.

When a Task sends a command to an I/O device, that device will spend some time doing it. To optimize the CPU use for as many Tasks as possible, a Task freezing technique is adopted when it starts an I/O procedure and is waiting data from the device; this characteristics is called **preemption**. Preemption allows other Tasks in ready status to use the CPU; when the device ends its job and data are ready to be transferred, it notifies its state using a new signal called **Interrupt** [9,13,1,14].

The Interrupt is usually managed by an integrated circuit, which sends the relevant information to the CPU. The Interrupt signals from each device are received by some devoted pins of the CPU, the **Interrupt request** (or IRQ) pins; the CPU decides which device to serve first on the basis of a priority mechanism and the right **Interrupt Service Routine** (ISR) takes care of the event. Then the Task which started the I/O procedure is waked up as soon as possible to complete its job and to free the resources it is using.

This procedure is sketched in Fig. 2. T_L is the **Interrupt latency time**; it is an important parameter, since it allows evaluating the responsiveness of a real-time system to an event.

Real-time software architectures can be divided according to the complexity and the speed of response guaranteed to the Tasks. In the following sections a brief survey of the two principal software architectures is presented; they both allow to manage the interaction with I/O ports and with the user, showing the right compromise between simplicity and adequacy of the components and real-time characteristics.

Round-robin with Interrupt The ISR of each I/O device takes part in the signal and protocol handling and reserves the data exchange service; an infinite loop Task looks after the requesting devices [13].

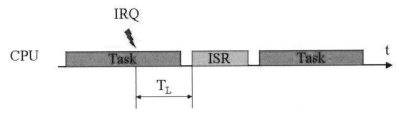

Fig. 2. Interrupt mechanism.

Note that the failure in one of the devices could lock the Task for a long time. Furthermore, the global variables accessible by the ISRs and by the Task must be protected; in fact, the ISRs could modify these variables in any instant and could give back to the Task a different environment without the Task knowing it.

Real-Time Operating System Maximum flexibility, but also maximum complexity, can be obtained using a real-time operating system (**RTOS**). ISRs and Tasks can be executed in parallel and according to fixed priorities; a software component called **scheduler** intervenes when particular "events" occur in the system, to switch the CPU control from a Task to a new one. This is the so-called **processes model** [14], in which an ISR or a Task represents a process, with all its data, its executable code and the system identification parameters. Each process carries out its own job sequentially and, virtually, disposes of a dedicated CPU; whereas actually, the only CPU in the system (in the mono-processor systems) is shared on time basis between all processes in execution.

2.3   Prototyping Systems for Robotics

Robotic systems are composed by several mechanical parts moved by means of electrical drives. In particular, a robot has several arms and joints arranged in a complex kinematic chain; the entire structure is moved combining the motion of each joint. A supervision/control system takes care of the activation "rules" for each drive so that the articulated structure carries out the desired tasks [12].

Command signals from the computing system to the **actuators** are the result of appropriate computation on data coming from **sensors**, the so-called **feedback control**. This modality is critical from a computational point of view because it must be characterized by a reasonable knowledge of execution times. In such cases the real-time requirements play an important role.

In this chapter our interest will be concentrated on these real-time systems, which represent a particularly complex branch of mechatronics, and on their prototyping.

The supervision of these systems requires the execution of several activities with hard or soft real-time issues; a list of these activities can be the following:

- position control of single joints,

- trajectory planning for coordinate movements,
- enabling phases and emergency control,
- man-machine interface.

At a higher task level, with respect to axis control, it is necessary to decide how a desired tip movement in the working space can be obtained and to send correct guidelines to each motor. Movements must take into consideration all the enabling requirements of the individual components and the safety for the operators and the machine. Anomalous behaviors, like unexpected collisions, motor current overloads or joint limit switch activations, need to be detected as soon as possible.

The system could also provide some sort of man-machine interaction: the operator may need a simple graphical interface to configure the functionalities of the manipulator tasks or use some more complex integrated diagnostic tool.

Therefore there are many complex and concomitant Tasks to execute; a single computer can manage all of them or it could be appropriate to devote a computer to the low level machine handling, leaving the soft-real-time Tasks to a separate one.

In the following section such architectures are briefly described.

Hardware architectures When prototyping is concerned, i.e. when systems usually do not work in "extreme" conditions and with heavy workloads, the set of available hardware components can include general purpose devices. In the last years the trend has been to use commercial-off-the-shelf (COTS) components due to their low costs and tested reliability.

The Host PC must be able to run the development tools, and to perform simulation processes of various kind, which is the most important assignment of a prototyping system.

The Target machine is often less powerful, but equipped with I/O boards to interact with the plant. Conversion speed of the on-board electronics, the communication bus between CPU and boards and Interrupt latency are typical bottlenecks for the Target.

Software architectures In the context of control for robotic systems, Tasks can be classified as **synchronous** (or **periodic**) and **asynchronous** (or **aperiodic**). A reliable internal mechanism to provide a timing base for synchronous events is needed. The asynchronous events occuring during the normal activities are dealt with in the spare time between the synchronous events.

Clock-driven architectures [9] are ideal candidates for this type of Tasks: the presence of periodic Tasks T_i, having well defined real-time execution characteristics is contemplated. An interrupt related to clock signals wakes up the scheduler according to the period of each Task. A Task T_i is defined by two parameters: the period P_i and the execution time e_i; it is assumed that T_i finishes its job before the end of its period, to guarantee the execution in the next period. The P_i can be different if the arrangement of T_i is done according to an hyperperiod equal to the least common multiple between all P_i. The Tasks scheduling must be arranged to allow the execution of all Tasks, according to their periods, into a unique hyper-period which will be repeated indefinitely.

This type of scheduling gives origin to inactive intervals, during which the aperiodic Tasks can be executed; these Tasks have soft-real-time characteristics and can treat "normal" situations.

There are also **sporadic** Tasks, which are usually devoted to react to unexpected events with hard real-time characteristics.

The clock-driven architecture can be implemented using both round-robin with interrupt and RTOS.

2.4   Prototyping Tools

The CAD environment allows representing the manipulator kinematics and dynamics by some sort of formalism. Simulink and Matlab from The Mathworks, Inc. are CAD tools widely used in research and design of control systems. Simulink allows assembling system parts according to a graphical block formalism. In Simulink it is possible to include event-driven process logic using the Stateflow tool; this instrument is based on **finite state machine** theory. A Stateflow diagram is composed by blocks representing states, and the simulator passes from one to another when some specified event happens. These events are associated to oriented edges linking the state blocks and labels specifying conditions and, possibly, actions. Both Mealy and Moore paradigms (actions associated to transitions and action associated to states, respectively) are supported. A Stateflow diagram included in a Simulink diagram can implement conditions and constraints on the execution of the overall simulation.

Automatic code generation Crucial to prototyping is the implementation of control and supervision algorithms on the actual controller: it is necessary to translate the block formalism into a high level language, usually C or C++. In order to obtain a Target processor executable it is necessary to perform program compilation by a **cross-compiler** residing on the Host PC. The program is then transferred and run on the Target PC using software tools resident on the Host, able to manage, monitor and in case also debug the testing progression.

The real-time software architectures described above are the starting point for building the control structure; to reduce the error possibilities and to cut the proto-typing process time, automatic code generation can be used. This process is called **rapid prototyping**: Real-Time Workshop (RTW) and Stateflow Coder are the tools which translate each block and the finite state machines in a programming language specified by the user, usually in C. There are also some rules to define how to code block relations and organization.

This last characteristic is interesting for real-time programmers because it gives the possibility to choose the resulting software architecture. Two architectural models, already described, are available:

- round-robin with Interrupt model,

- processes model based on RTOS.

Both architectures can manage multitasking; in the following sections two implementation example are described using pseudocode.

Round-robin with Interrupt

```
main() {
Initialization (including installation of rtISR as an interrupt
service routine, ISR, for a real-time clock)
While(time < final time)
Background task
EndWhile
Mask interrupts (Disable rtISR from executing)
Complete any background tasks
Shutdown
}

rtISR() {
Check for interrupt overflow
Enable "rtISR" interrupt
Update outputs and discrete states (tid=0) and log data
Update continuous states
For i=1:NumTasks
If (hit in task i)
Update outputs and discrete states (tid=i)
EndIf
EndFor
}
```

The *rtISR* procedure is executed when an Interrupt is generated by a clock, with a cadence equal to the fastest sampling time present in the model. Its structure is similar to the simulation mechanism: output update, discrete states update, continuous states integration (if present, the ISR execution period equals the integration step). Multitasking is built imposing multiple sampling times with respect to the basic Task ($tid=0$), so that for each Interrupt cycle the states having the sampling time tick in that instant are updated. During *rtISR* inactivity period a Background Task with non-real-time jobs is executed. The whole mechanism is started by the *main* routine which organizes the real-time clock, the ISR and the Background execution cycle; the same routine ends the execution, masking the Interrupt signal and completing the Background.

Multiprocess with RTOS primitives

```
main() {
Initialization
Start task "tBaseRate".
Start task "tSubRate".

Start clock that does a "semGive" on a clockSem semaphore.
Wait on "model-running" semaphore.
Shutdown
}

tSubRate(subTaskSem,i) {
Loop:
Wait on semaphore subTaskSem.
```

```
Update outputs and discrete states (tid=i)
EndLoop
}
tBaseRate() {
MainLoop:
If clockSem already "given", then error out due to overflow.
Wait on clockSem
For i=1:NumTasks
If (hit in task i)
If task i is currently executing, then error out due to overflow.
Do a "semGive" on subTaskSem for task i.
EndIf
EndFor
Update outputs and discrete states (tid=0) and log data
Update continuous states
EndMainLoop
}
```

In this case the model is executed using some typical RTOS primitive: processes creation and start with fixed priority, and Task synchronization by means of semaphores. The *main* procedure creates a *tBaseRate* process having the highest priority, i.e. waked up by the fastest clock period of the model. More *tSubRate* processes with multiple sampling time with respect to *tBaseRate* are created, each one with decreasing priority and a more relaxed sampling time. At each activation *tBaseRate* checks and unlocks by means of semaphores the *tSubRate*, which must be executed in the same sampling time. However, since *tBaseRate* has the highest priority, it continues to execute its jobs, preventing other process executions and completing the elaboration of the fastest part of the model. When it finishes, the previously unlocked *tSubRate* procedures start to execute their jobs using the CPU on the basis of their priorities.

It should be noted that one of the key concepts which grants the multitasking execution with hard real-time requirements in these control software architectures is the relation between the sampling times of each Task: the base sampling time is decided on the basis of the requirements of the most critical Task, the other Tasks being executed with multiple sampling times with respect to the base sampling time.

Host-Target communications The Host machine is usually supervised by a general purpose operating system with graphical interfaces, allowing a simple and direct interaction with the user in the design and development phases and for the management of Target-plant interaction. No real-time requirements are necessary: the user prepares its tasks off-line, sets the Target execution, and, at the end, analyzes the obtained data.

Recent technologies allow data exchange between Host and Target using TCP/IP on Ethernet or RS232 protocols. The Target machine is bounded by hard and soft real-time requirements and cannot interrupt its Tasks in a given instant. The Host requires asynchronous mode interaction and the Target reacts as soon as possible, respecting the highest priorities of real-time Tasks. These facts motivate the two fundamental techniques for data exchange:

- "on-the-fly" transfer, in which the Target tries to communicate with the Host during the real-time Task execution,

- transfer at the end of the current test session, in which, before the tests starts, the Host asks the Target to collect data, the Target executes the test memorizing relevant data, and when each real-time Task ends its jobs and frees the CPU, it transfers the data to the Host.

The first technique does not guarantee that the Host receives all the data related to the real-time signals: the Target could be in a busy state executing a real-time procedure, and it could not manage the communication exchange. This fact can originate an incorrect reconstruction of the observed signals due to data incompleteness.

When Round-robin with Interrupt is used, the transfer job can be executed as a background Task when the ISR does not run; however the ISR can interrupt the communication in any instant, provoking partially data losses. In the RTOS architecture, data collection and transfer to the Host is usually executed by a low-priority process, which is interrupted by the scheduler when a real-time process has to be executed; obviously, data losses can occur in this case too. Actually this low-priority process can be a web server, and it can manage queries coming from clients all over the net.

The data transfer at the end of the test, according to the second technique, makes possible a correct reconstruction of all signals. This job is not particularly expensive in terms of CPU time for the real-time systems; the Task which manages data storage uses the CPU for brief time intervals, and if the workload of the real-time system is not critical, there is a high probability to complete the job in time. Unlike the first technique, it is possible to correctly reconstruct the acquired signals, paying the price of a retard in the data exchange with the Host, and of growing memory needs.

3   The Prototyping Environment

In this section, the software and hardware architecture of a fast prototyping environment developed at the Robotic Laboratory of Politecnico di Torino will be described. It relies on a round-robin with Interrupt architecture and is implemented on a DSP based controller, managed through a Matlab toolbox running on the Host PC.

In Section 4 some experiments and results obtained with this environment will be illustrated.

3.1   The Robotic System

The experiments were performed on a double-arm planar manipulator with revolute vertical axis joints, sketched in Fig. 3.

Two **brushless NSK Megatorque** direct-drive (i.e. without gearboxes) motors move the joints. The maximum extension of the links ($L_1 + L_2$) is about 0.7 m, the angular limits are ± 2.15 rad for both joints, and the tip height moves parallel to the horizontal plane at a distance of 0.45 m; joint angular positions are measured by internal **resolvers**.

The two motors are actuated by power drives, which take care of the various and complex functions of these motors and look after the signals coming from the

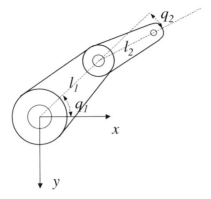

Fig. 3. Diagram of the double-link planar manipulator used for testing.

resolvers. The drive communication system deals, in particular, with some of the main features that are basic for control, such as digital input/output signals interchange, application of analog command inputs, and decoding of position information from sensors.

The drive cabinets contain power electronics for the PWM of the motors, and a card devoted to transform analog signals from resolver into digital signals of shaft encoder type, based on a 16 bit microprocessor.

The analog signals coming from the controller are interpreted as torque or velocity reference commands to be applied to the motors, according to the two available control modes: **Torque Mode** and **Velocity Mode**. On the basis of the resolver signals, a current loop is closed to control the torque in the first case, whereas an additional velocity loop is added in the second mode. The default mode used to test different types of control algorithms is the Torque mode.

The inner current loop parameters are fixed, and the actuator model can be approximated by a simple proportional gain $K_{v\tau}$ between the input command voltage, V_m, and the torque τ_m supplied by the motor

$$\tau_m = K_{v\tau} V_m \tag{1}$$

The optional Velocity Mode is useful in emergency situations, when the user needs to instantly arrest the manipulator motion, pushing the STOP button: a digital input linked to the button activates the velocity control loop, imposing zero velocity reference. The stopping phase will be executed as specified by the internal velocity control algorithm.

The overall plant and the controller can be modelled as in the diagram of Fig. 4, that shows how the controller receives encoders signals and gives back voltage signals in mV, proportional to required command torques.

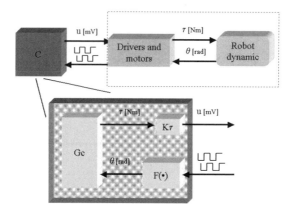

Fig. 4. IMI-ODSP model.

3.2   The Control System Architecture

The original control system has been replaced by a new one, in which the components for real-time interaction are grouped in a modular industrial standard rack.

This control system environment, called **OpenDSP**, has been developed by the Mechatronics Laboratory of the Politecnico di Torino and consists of a DSP board and a programmable input/output board. A PLD (Programmable Logic Device) on the latter board allows configuring via software the digital and analog inputs and outputs, and preprocessing these signals in a customized way, before they reach the converters or the DSP. Field interfacing is obtained by means of user customizable boards, packaged with the I/O board and the DSP board in the same rack. The real-time control requirements are guaranteed by the presence of a link between the I/O and the DSP boards based on a proprietary bus (called the OpenDSP bus).

The system is linked via enhanced parallel port (EPP) protocol to a desktop PC, working as a Host, and by connections to each axis interface.

A Matlab environment with Simulink runs on the Host PC. The OpenDSP system includes a new toolbox for Matlab called **MatDSP**, which allows Matlab-code interaction with the DSP. MatDSP too has been developed by the Mechatronics Laboratory of the Politecnico di Torino.

MatDSP makes possible, among other functions, to read and/or change any variable processed by the DSP. For example, the parameters of a control algorithm can be changed "on fly" in a single sampling time in order to guarantee a coherent switch to the new configuration (synchronous mode); or different variables, at user's choice, can be monitored without requiring a more stringent "sample by sample" acquisition (asynchronous mode). It is possible to monitor the real-time variables and the drives status flags, to scope and acquire signals and make any type of mathematical operation on them. A control algorithm written in C can be compiled, downloaded and started/paused on the DSP.

Simple graphic user interfaces have been built in the Matlab environment using the GUIDE tool, to simplify testing and management of signals exchanged with the drives. The MatDSP commands have been hidden by a logic construction, grouping the signals in high level functions rather than using them to perform single hardware operations.

For example, a large number of cross-controls is needed to guarantee the correct and safe sequence of operations to enable and start the control task; this would oblige the user to read and change several variables using the primitive statements provided by the MatDSP toolbox. On the contrary, hiding the MatDSP commands under these GUIs allows the user to concentrate on new experiments. An example of one of these GUIs is shown in Fig. 5.

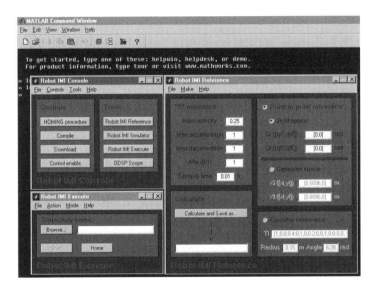

Fig. 5. A GUI example for the double arm manipulator supervision.

Three tools are available to the user: the first one, called **IMIConsole**, is a GUI panel to perform the homing procedure, to prepare and to enable a control algorithm chosen from a list; it is the entry point for the normal interaction with the control system.

The second tool, called **IMIExecute**, is a GUI panel that allows selecting and executing, in single or cyclic mode, a previously planned trajectory and make a home return to the zero position. This GUI shares the same data base of IMIConsole tool to ensure appropriate and safe operations.

The third tool, called **IMIReference**, does not interact with the system as it is not related to the MatDSP toolbox, unlike the IMIConsole and the IMIExecute GUIs. It is used to generate some simple, basic reference functions, such as joint or Cartesian point-to-point moves or circular trajectories, and save them in a MAT file.

From the IMIConsole panel it is possible to open the IMIExecute or the IMIReference GUIs and load a Simulink model of the robot to simulate the planned trajectories before executing them on the real plant. The user can test and change the structure or the parameters of a control algorithm until a satisfactory response is reached.

The designer can now translate the algorithm in C code, compile and download it using the GUIs, impose the same trajectory used in simulation, and enable the robot to execute it. If the experiment is satisfactory the prototyping session ends, otherwise the procedure is repeated with a refined Simulink model or with a new control algorithm.

3.3   The OpenDSP Software Architecture

The OpenDSP real-time software relies on a round-robin with Interrupt architecture. When the system is initialized, a main function, *Main.c*, calls some sub-functions which configure the system on the basis of a group of parameters, some of which fixed and other ones assigned by the user. Then, in an infinite loop two other sub-functions are called in turn: the first one, called *Monitor*, takes care of data exchange between the Matlab environment and the DSP; the second one, called *UserBackground*, allows executing a user code at a lower priority level, which interprets and executes the Matlab commands and interacts with the drives' logic. Both sub-functions have no hard-real-time requirements and can be interrupted when the periodical axis control function, written by the user, starts.

The whole user code is divided in sections and hosted in a file on the basis of a C written **template**; no automatic code generation has yet been implemented in this prototyping system. The initial section, the *UserInit*, contains the code to initialize the customizable characteristics of the system and the starting settings of axis control functions; it is executed one time, when the code downloaded to the DSP is launched. The variables, which must be available in the Matlab workspace, are declared and initialized within this function.

User writes in the subsequent *UserISR_INT2* section the control algorithm code and all the functions useful to close the loop: sensors reading, position reference management and command application. *UserISR_INT2* is executed every control sampling time according to the following procedure:

- a timer sends a signal for Start Of Conversion (SOC) to the input and output converters (ADCs and DACs);
- when the conversion ends, a signal for End Of Conversion (EOC) returns, and the DSP stops the current job, i.e., one of the Monitor or UserBackground functions; note that a sampling time delay is inserted by the system in the model of the plant, since the DAC uses the command computed in the previous step;
- *UserISR_INT2* is executed, and afterwards the DSP returns to the suspended job.

The sequence assumes that the control algorithm computation ends before the next EOC signal, to allow the execution of portions of non real-time jobs, too.

The template is ended by the UserBackground function, that contains the code executed by the DSP when the Monitor and *UserISR_INT2* functions are inactive. As previously said, this code interprets the commands coming from Matlab and pass them to the DSP environment by means of the Monitor function.

To summarize, the open architecture of this system has allowed to configure five sections of the whole structure.

- The hardware interface toward the plant, using custom electronics built on a standard development field module to be mechanically compliant with the rack and the stackthrough structure.
- The logical interface between DSP and field modules, managed by the PLD firmware. Starting from a general architecture, the PLD **user part** is initialized with suitable logic circuits devoted to group and convert signals from and to the field module in registers, or to close faster loops (in microseconds).
- The data base structure of the real-time signals, built in the form of registers and channel manageable by suitable macros in a pre-structured C header file.
- The Background routine that manages the communication between Host and DSP, and the ISR routine to control the axes, starting from a general and strongly organized C template.
- The asynchronous communication between Matlab user and plant by means of a graphic user interface giving a logical and easier interpretation of the plant functionalities.

4   Description of a Test Case: Prototyping a Model-Based Compensation of Nonlinear Joint Friction

The model of the manipulator under study [2], [3] can be described by the following second-order nonlinear differential equation:

$$M(q)\ddot{q} + C(q,\dot{q})\dot{q} + \tau_f(q,\dot{q}) = \tau_m \tag{2}$$

where q, $\dot{q}$, and $\ddot{q}$ are the vectors of joint angles, angular velocities and angular accelerations, $M(q)$ is the configuration-dependent inertia matrix, including both links and motors inertia, $C\dot{q}$ is the term containing Coriolis and centrifugal torques, τ_f is the friction torque vector, and τ_m is the command torque vector. No gravity term is present, since the manipulator moves in a horizontal plane. The electrical time-constants of the motors are not considered, as the inner current loop guarantees that they are much faster than the mechanical ones, and that, consequently, the relationship between the input voltage and the output torque is simply given by a known gain $K_{v\tau}$.

The determination of a proper model to describe the friction phenomena, whose effects are modelled in τ_f, and the identification of its parameters values have been performed by a series of appropriate tests, and executed by means of an appropriate C-based DSP code, developed within fixed templates. In particular, two different procedures have been applied to perform two different kinds of tests:

- open-loop tests (to estimate stiction and friction at high velocity), with the joints free to rotate;
- closed-loop tests (to estimate static friction at low velocity, and dynamic friction in the presliding phase), with the manipulator in the controlled configuration.

In particular, starting from the acquired joint position samples and the corresponding velocity values, computed using a simple digital filter, the friction torques have been indirectly derived by considering:

- in the open-loop tests:

$$\boldsymbol{\tau}_{m,k} = \boldsymbol{\tau}_{f,k} \tag{3}$$

where $\boldsymbol{\tau}_{m,k}$ and $\boldsymbol{\tau}_{f,k}$ are the k-th samples of the applied motor torques and of the joint friction torques, respectively;
- in the closed-loop tests at low velocity:

$$\boldsymbol{\tau}_f(\dot{\boldsymbol{q}}) + \boldsymbol{\tau}_{err} = \boldsymbol{\tau}_m - M(\boldsymbol{q})\ddot{\boldsymbol{q}} - C(\boldsymbol{q},\dot{\boldsymbol{q}}) \tag{4}$$

from the manipulator dynamic equation (2), where $\boldsymbol{\tau}_{err}$ is a torque vector that represents all modelling errors and measurement disturbances; such a term has been disregarded, repeating several times the same motion and filtering the measured data to extract the mean values.

Stiction (i.e. friction at zero velocity) has been estimated by tests in which each joint is set in a definite angular position, the drive is set in Torque Mode, and minimal torque increments are supplied in both clockwise (CW) and counterclockwise (CCW) directions. No joint motion is noticeable until the command torque reaches the maximum static friction value. When the joint starts to rotate, the current torque value is registered, and the procedure is repeated for various starting angular positions, to test the stiction dependency on the angular position of the joint.

Tests are executed by means of a DSP code based on a fixed template, modified just in the section relative to the control function, the UserISR_INT2. The command torque increments are supplied in open loop, directly from the user.

The test is executed in the Matlab environment using the IMIConsole GUI to compile and download the real-time code and to enable the axis drives; run-time changes of the command torque reference are allowed by the commands `MatDSPvariable(VarName, NewValue)` and `MatDSPupdate`. In particular, the last command lets all the real-time variables, modified by the user with the command `MatDSPvariable`, be refreshed in the same sampling time.

Finally, the mean stiction value is computed and used as the estimated stiction value.

The contribution of viscous friction at high velocity has been evaluated letting the joints rotate freely, and using the Torque Mode functionality to achieve a situation of dynamic equilibrium at constant velocity, in which the inertial torque is zero, and the friction torque can be assumed to be approximately equal to the command torque.

The DSP code necessary for these experiments is the same used to evaluate stiction, with the addition of the position measurement by means of the macro

`IOGP_FU1_READ_ENC_CURRENT(Channel)` and the acquisition data command, `Acquire()`, at the end of function UserISR_INT2.

This functionality offered by the system is configurable at run-time by the Matlab command `MatDSPAcquireConfig(params)`, choosing: i) which data are to be acquired, ii) data decimation parameters, and iii) the acquisition time interval. It is not an invasive operation for the control function, i.e., it does not cause the violation of the sampling time, because it is executed entirely in the DSP environment to avoid a slow data exchange with the PC. The Monitor function returns acquired data to Matlab environment, without real-time constraints, when the user invokes the command `MatDSPAcquireLoad()`. In the considered case, angular joint position values are acquired for each torque increment. A waiting time interval allows the end of the acceleration fluctuations, after which a two seconds acquisition is started. Angular velocity data are computed from the measured positions, for each joint and for each rotation direction, and for every velocity sample the corresponding friction torque is assumed equal to the command torque τ_m. The velocity data obtained have a lower bound value of about 2 rad/s, due to the sudden transition from stop to motion and viceversa.

Joint friction at low velocity has been then investigated by an experimental session performed with the manipulator in the controlled configuration. A simple PD control law is used to assign to each joint the position/velocity profile defined by the user, to properly collect data for the estimation of static friction at low velocity, and dynamic friction in the presliding phase. More code is added at the UserISR_INT2 to supply a micro-interpolation mechanism for the user profile, together with a section devoted to the position data processing needed by the PD algorithm. The IMIExecute GUI is used, together with the IMIConsole, to transfer the reference position vector to the DSP running code, which interpolates and executes the movement. The user provides the reference vector and the data acquisition request by means of the IMIExecute, and then, after a pre-positioning phase, the task is executed and a MAT file containing the acquired data is saved in a predefined directory.

On the basis of the acquired data, the well-known *LuGre* model [6], [11] has been considered to represent the friction torques on each joint of the manipulator. Such a model includes both a steady-state (*static*) friction curve, and the *dynamic* friction behavior during the presliding phase by means of a "bristle" model, according to the following equations:

$$\frac{\mathrm{d}z_i}{\mathrm{d}t} = \dot{q}_i - \frac{|\dot{q}_i|}{g_i(\dot{q}_i)}\sigma_{0,i}z_i \qquad (5)$$

$$\tau_{f,i} = \sigma_{0,i}z_i + \sigma_{1,i}\frac{\mathrm{d}z_i}{\mathrm{d}t} + f_i(\dot{q}_i) \qquad (6)$$

where z_i is a state variable representing the average bristle deflection for joint i, $\sigma_{0,i}$ and $\sigma_{1,i}$ are model parameters that are assumed to be constant, and functions $g_i(\dot{q}_i)$ and $f_i(\dot{q}_i)$ model the Stribeck effect and the viscous friction, respectively. For constant velocity, the steady-state friction torque is then given by:

$$\tau_{f,i_{ss}} = g_i(\dot{q}_i)\,\mathrm{sgn}(\dot{q}_i) + f_i(\dot{q}_i) \qquad (7)$$

Among the different parameterizations that can be used to describe $g_i(\dot{q}_i)$ and $f_i(\dot{q}_i)$, the following ones have been chosen because they fit well the acquired data:

$$g_i(\dot{q}_i) = \alpha_{0,i} + \alpha_{1,i}e^{-\frac{\dot{q}_i}{\dot{q}_{s1,i}}\operatorname{sgn}(\dot{q}_i)}$$

$$+\alpha_{2,i}\left(1 - e^{-\frac{\dot{q}_i}{\dot{q}_{s2,i}}\operatorname{sgn}(\dot{q}_i)}\right) \tag{8}$$

$$f_i(\dot{q}_i) = \alpha_{3,i}\dot{q}_i + \alpha_{4,i}\dot{q}_i^2 \tag{9}$$

The *static* parameters in (8) and (9) (i.e., the four $\alpha_{k,i}$'s for each joint, together with $\dot{q}_{s1,i}$ and $\dot{q}_{s2,i}$), have been estimated by considering tentative values between 0.1 and 0.3 rad/s for the exponential parameters $\dot{q}_{s1,i}$ and $\dot{q}_{s2,i}$ (on the basis of the acquired data), and applying a least square algorithm to a linearized expression of (7)-(9) to estimate the α's parameters for each joint. By some iterations, the values reported in Table 1 have been obtained.

Table 1. Estimated static parameters of the LuGre friction model.

	Joint 1 $\omega > 0$	Joint 1 $\omega < 0$	Joint 2 $\omega > 0$	Joint 2 $\omega < 0$
α_0	40.854	−46.473	17.837	3.408
α_1	−32.454	53.873	−14.837	−0.408
α_2	−31.233	55.738	−14.998	−0.635
α_3	−0.760	−0.293	−0.156	−0.104
α_4	−0.262	0.177	−0.050	0.036
$\dot{q}_{s1}$	0.19	0.14	0.2	0.3
$\dot{q}_{s2}$	0.17	0.15	0.19	0.1

Figure 6 shows the resulting steady-state friction torque together with the acquired data for the first joint (for positive and negative velocities). Similar results have been obtained for the second joint.

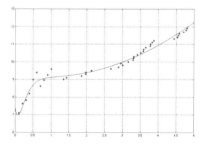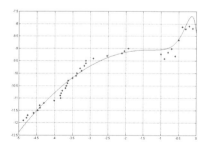

Fig. 6. Friction torque (Nm) on joint 1 for positive and negative velocities (rad/s).

Following a procedure similar to the one proposed in [5], under some *pre-sliding* assumptions, the *dynamic* friction parameters $\sigma_{0,i}$ and $\sigma_{1,i}$ have been estimated, computing z_i by integrating equation (5) from joint position measures, acquired during an appropriate motion of each joint. In particular, a slowly growing torque ramp has been applied to each joint to estimate $\sigma_{0,i}$, and a torque step to estimate $\sigma_{1,i}$ (see [5] for details). A good approximation of such parameters could be obtained only starting from high precision joint position measurements. Since in our case the encoder signal is decoded with a resolution of only $2\pi/76800$ rad, a subsequent model validation phase has been performed, by comparing the real robot behavior with the results of some simulation tests, carried out by a Simulink model, which can be directly run from the IMIConsole GUI. Some adjustments of the estimated values of the dynamic friction parameters have been allowed by this procedure, but some additional investigation will be necessary. The currently estimated values are reported in Table 2.

Table 2. Estimated dynamic parameters of the LuGre friction model.

	Joint 1 $\omega > 0$	Joint 1 $\omega < 0$	Joint 2 $\omega > 0$	Joint 2 $\omega < 0$
σ_0	55500	26000	12600	12600
σ_1	1000	800	70	70

Preliminary tests have been performed to evaluate the improvements that can be obtained from the control point of view by friction compensation. The applied inverse dynamic control scheme, including only *static* friction compensation (as the currently available dynamic friction estimation is not yet satisfactory), is of the following type:

$$\boldsymbol{\tau}_m = \boldsymbol{M}(\boldsymbol{q})(\ddot{\boldsymbol{q}}_r - \boldsymbol{v}_c) + \boldsymbol{C}(\boldsymbol{q}, \dot{\boldsymbol{q}})\dot{\boldsymbol{q}} + \hat{\boldsymbol{\tau}}_f(\dot{\boldsymbol{q}}) \qquad (10)$$

where $\ddot{\boldsymbol{q}}_r$ is the joint acceleration reference vector, $\hat{\boldsymbol{\tau}}_f(\dot{\boldsymbol{q}})$ is the estimated friction torque vector, and a PD control algorithm has been considered to define the outer loop law $\boldsymbol{v}_c$. The corresponding DSP code is very similar to the one used in the low velocity friction estimation tests; sub-sections, containing the robot inverse dynamics and an high-order polynomial function, approximating the estimated static friction model, have been simply added within the UserISR_INT2.

Figure 7 shows the time history of the resulting position error of the first joint, for a circular Cartesian reference trajectory, defined by means of the IMIReference GUI, when $\hat{\boldsymbol{\tau}}_f(\dot{\boldsymbol{q}}) = 0$ is considered, i.e. without any friction compensation, and when $\hat{\boldsymbol{\tau}}_f(\dot{\boldsymbol{q}})$ corresponds to the estimated steady-state friction curve. As Figure 7 shows, even though only static friction has been compensated, a significant error reduction has been obtained; similar results have been achieved for the second joint, too.

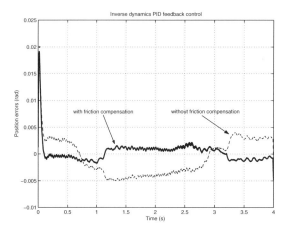

Fig. 7. Inverse dynamic control without and with friction compensation: position error on joint 1.

5 Conclusion

Rapid prototyping systems are used to speed up the development of a final product. In industrial robotics these systems have real-time requirements fulfilled by means of architectures which allow the designer to concentrate his work on the prototype development.

At the Politecnico di Torino a solution based on the well known round-robin with Interrupt software architecture, the OpenDSP system, has been proposed. The hardware architecture is based on a Host-Target solution, with the Host running Matlab and a dedicated toolbox to manage the Target DSP board.

Some experiments devoted to model and control the friction phenomena have been presented, demonstrating a right tradeoff between usability and efficiency of the OpenDSP system.

In perspective, an advanced prototyping architecture based on standard real-time operating systems will be investigated and implemented, aimed at providing an extended environment, with increased interaction capabilities between data acquisition, model analysis and control design.

Particular attention will be devoted to integrate discrete state/event transitions and continuous control, extended use of graphical modelling (under Simulink), automatic code generation, vision and force sensors integration. From this point of view an interesting real-time operating system is Linux with RTAI patch. RTAI (Real-Time Application Interfaces) [15] provides Linux with real-time features (appropriate syscalls, RT scheduling, reduced latency, etc.) and allows, in its present development state, the automatic code generation from Matlab Real-Time Workshop.

Acknowledgement

This work has been co-funded by the Italian Space Agency (ASI). A particular note of thanks goes to CSPP-LIM, Politecnico di Torino, for technical support.

References

1. K. Arnold, *Embedded Controller Hardware Design*, LLH Technology Publishing, 2001.
2. B. Bona, M. Indri, and N. Smaldone, "Open system real-time architecture and software design for robot control," *Proc. of 2001 IEEE/ASME Int. Conf. on Advanced Intelligent Mechatronics*, pp. 349–354, 2001.
3. B. Bona, M. Indri, and N. Smaldone, "An experimental setup for modelling, simulation and fast prototyping of mechanical arms," *Proc. of 12th IEEE Conf. on Computer-Aided Control Systems Design*, pp. 207–212, 2002.
4. C.J. Boots and G.P. Kurpis, *The New IEEE Standard Dictionary of Electrical and Electronic Terms [Including Abstracts of All Current IEEE Standards]*, 5th Ed., IEEE, 1993.
5. C. Canudas de Wit and P. Lischinsky, "Adaptive friction compensation with partially known dynamic friction model," *Int. J. of Adaptive Control and Signal Processing*, vol. 11, pp. 65–80, 1997.
6. C. Canudas de Wit, H. Olsson, K.J. Åström, and P. Lischinsky, "A new model for control of systems with friction," *IEEE Trans. on Automatic Control*, vol. 40, pp. 419–425, 1995.
7. H. Hanselmann, "Automotive control: From concepts to experiments to product," *Proc. of 9th IEEE Int. Symp. on Computer-Aided Control Systems Design*, pp. 129–134, 1996.
8. F. Kordon and Luqi, "An introduction to rapid system prototyping," *IEEE Trans. on Software Engineering*, vol. 28, pp. 817–821, 2002.
9. J.W.S. Liu, *Real-Time Systems*, Prentice-Hall, 2000.
10. G. Mason, A. Pongpunwattana, and M. Berg, "Design and modeling of a flexible test bed for use in control system analysis and verification," *Mechatronics*, vol. 12, pp. 891–904, 2002.
11. H. Olsson, K.J. Åström, C. Canudas de Wit, M. Gäfvert, and P. Lischinsky, "Friction models and friction compensation," *European J. of Control*, vol. 4, pp. 176–195, 1998.
12. L. Sciavicco and B. Siciliano, *Modelling and Control of Robot Manipulators*, 2nd Ed., Springer Verlag, 2000.
13. D.E. Simon, *An Embedded Software Primer*, Addison Wesley, 2001.
14. A.S. Tanenbaum, *Modern Operating Systems*, 2nd Ed., Prentice-Hall, 2001.
15. http://www.aero.polimi.it/~rtai/.

Real-Time Visual Tracking of 3D Objects

Fabrizio Caccavale[1], Vincenzo Lippiello[2], Bruno Siciliano[2], and Luigi Villani[2]

[1] Dipartimento di Ingegneria e Fisica dell'Ambiente
  Università della Basilicata
  Contrada Macchia Romana, 85100 Potenza, Italy
  caccavale@unibas.it
[2] PRISMA Lab
  Dipartimento di Informatica e Sistemistica
  Università di Napoli Federico II
  Via Claudio 21, 80125 Napoli, Italy
  <vincenzo.lippiello,bruno.siciliano,luigi.villani>@unina.it
  http://www.prisma.unina.it

Abstract. The use of visual sensors may have high impact in applications where it is required to measure the pose (position and orientation) and the visual features of objects moving in unstructured environments. In robotics, the measurements provided by video cameras can be directly used to perform closed-loop control of the robot end-effector pose. In this chapter the problem of real-time estimation of the position and orientation of a moving object using a fixed stereo camera system is considered. An approach based on the use of the Extended Kalman Filter (EKF) combined with a 3D representation of the objects geometry based on Binary Space Partition (BSP) trees is illustrated. The performance of the proposed visual tracking algorithm is experimentally tested in the case of an object moving in the visible space of a fixed stereo camera system.

1 Introduction

In the last decade, research on visual sensing has received a new impulse because digital signal processing hardware with high computational capability is becoming available at low cost. In fact, visual sensors offer the possibility to extract a great variety of information from a scene in a noninvasive manner. This information can be used by automatic systems either at high level, e.g., for inspection, recognition and planning tasks, and at low-level, e.g., for autonomous guidance of vehicles, real-time control in scarcely structured environments.

In robotics, the measurements provided by video cameras can be directly used to perform closed-loop position/orientation control of the robot end effector, usually denoted as *visual servoing* control [11]. In this framework, two different approaches have been developed. The first approach is the position-based visual servoing, which defines the tracking error in the Euclidean space and requires the estimation of the position and orientation of a target object with respect to a reference frame [27]. The second approach is the image-based visual servoing, which defines a tracking error directly in the image space of the cameras, thus avoiding accurate calibration of the vision system [5,10]. Hybrid methods using position-based visual servoing to control

B. Siciliano et al. (Eds.): Advances in Control of Articulated and Mobile Robots, STAR 10, pp. 125–151, 2004.
© Springer-Verlag Berlin Heidelberg 2004

certain degrees of freedom and image-based visual servoing to control the remaining degrees of freedom can be adopted [7,19]. More recently, vision measurements have been used in combination with force measurements to develop control strategies aimed at improving the robot performance for the execution of tasks in scarcely structured environments [2].

In the case of position-based visual servo, computationally efficient techniques for *visual tracking* of the pose (position and orientation) of the target object must be adopted. One of the major problems to cope with is represented by the noise and disturbances affecting the visual measurements, due to temporal and spatial sampling and quantization of the image signal, lens distortion, etc., which may produce large pose estimation errors. The use of the Extended Kalman Filter (EKF) may improve the accuracy and speed of the visual tracking algorithm [15,25,21,17].

In fact, Kalman filtering offers many advantages over other pose estimation methods [1,9,28], e.g., implicit solution of photogrammetric equations with recursive implementation, temporal filtering, ability to change the measurement set during the operation. Moreover, the statistical properties of Kalman filter may be tuned to those of the image measurements noise of the particular vision system. Last but not least, the prediction capability of the filter allows setting up a dynamic windowing technique of the image plane which may sensibly reduce image processing time. Applications of Kalman filter in machine vision range from visual tracking of objects with many internal degrees of freedom [20], to automatic grasp planning [13] as well as pose and size parameters estimation of objects with partially known geometry [14].

A widely adopted strategy for object pose computation is based on the recognition of some geometric features of the object, such as edges and corners, from a camera image. In particular, the extraction of a suitable number of corners (feature points) allows computing the pose by using a simple point CAD model of the object [27,13]. In principle, the accuracy of the estimate increases with the number of the available feature points, at the expense of the computation time. However, when Kalman filter is adopted, it has been shown that the best achievable accuracy that can be obtained using all the available points is quite the same as that obtained using a number of five or six feature points, if properly chosen [25].

The choice of the optimal feature points can be performed by using suitable selection algorithms, whose complexity grows at factorial rate with the number of the available points [6,12]. Hence, to reduce the computational burden in the presence of a large number of feature points, it is crucial to perform a pre-selection, e.g., by eliminating all the points that, in a given object pose, are occluded with respect to the camera [24,8].

In this chapter, the EKF is adopted for real-time visual tracking of an object in the Euclidean space. In order to reduce computational time, a new pre-selection algorithm of the feature points is proposed, based on the selection of all the points that are visible to the camera at a given sample time. This algorithm exhibits a complexity which grows linearly, thanks to the use of Binary Space Partitioning (BSP) tree for object geometric representation [4]. In detail, the prediction of the object pose provided by the Kalman filter is used to drive a visit algorithm of the

BSP tree which allows identifying all the feature points that are visible at the next sample time. After the pre-selection, a dynamic windowing algorithm and an optimal point selection algorithm are adopted to find the windows of the image plane to be processed and input to the Kalman filter.

The proposed pre-selection algorithm can be used also in the case of objects and obstacles with interposing parts. Differently from other algorithms (see [12] and references therein), this method allows recognizing all the points of the surfaces of the objects which are hidden to the camera or occluded by some other objects or obstacles of known geometry [18].

The effectiveness of the proposed approach is tested in experimental case studies where the position and orientation of an object carried by a robot manipulator is tracked using both one fixed camera and a stereo system composed by two fixed cameras.

The chapter is organized as follows. In Section 2 the pin-hole model of the cameras is introduced and the photogrammetric equations are derived. The model used for object motion and the equations of the EKF are presented in Section 3. In Section 4 a BSP tree is derived form a CAD geometric model of an object. Section 5 is devoted to illustrate the pre-selection algorithm and an optimal point selection technique, based on dynamic windowing and quality indices. The whole estimation procedure is analyzed in Section 6. Finally, the experimental set up and the experimental tests are described in Section 7, while Section 8 presents some concluding remarks and open problems. Details on the derivation of EKF equations are reported in the Appendix.

2   Modelling

Consider system of n video cameras fixed with respect to a base coordinate frame $O\text{--}xyz$ and the pin-hole model of camera i (see Fig. 1). Let $O_{ci}\text{--}x_{ci}y_{ci}z_{ci}$ be a frame attached to the camera (camera frame), with the z_{ci}-axis aligned to the optical axis and the origin in the optical center. In the following, a superscript will be used to denote the reference frame of a variable, when different from the base frame.

For each camera, the sensor plane is parallel to the $x_{ci}y_{ci}$–plane at a distance $-f_e^{ci}$ along the z_{ci}–axis, where f_e^{ci} is the effective focal length of the camera lens, which may be different from the nominal focal length f^{ci}. The image plane is parallel to the $x_{ci}y_{ci}$–plane at a distance f_e^{ci} along the z_{ci}–axis. The intersection of the optical axis with the image plane defines the principal optic point O'_{ci}, which is the origin of the image frame $O'_{ci}\text{--}u_{ci}v_{ci}$ whose axes u_{ci} and v_{ci} are taken parallel to the axes x_{ci} and y_{ci}, respectively.

A point P with coordinates $p^{ci} = [\,x^{ci}\ \ y^{ci}\ \ z^{ci}\,]^{\mathrm{T}}$ in the i-th camera frame is projected onto the point of the image plane whose coordinates can be computed with the equation

$$\begin{bmatrix} u^{ci} \\ v^{ci} \end{bmatrix} = \frac{f_e^{ci}}{z^{ci}} \begin{bmatrix} x^{ci} \\ y^{ci} \end{bmatrix} \tag{1}$$

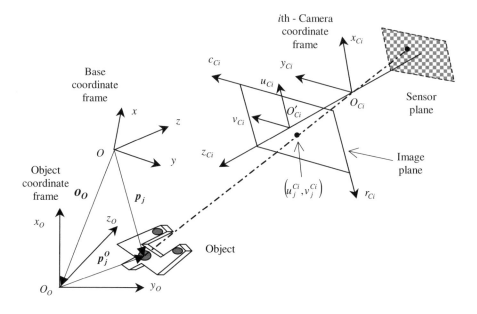

Fig. 1. Reference frames for the i-th camera and the object using the pin-hole model.

which is known as perspective transformation. A spatial sampling can be applied to the image plane by expressing the coordinates in terms of number of pixels as

$$\begin{bmatrix} r^{ci} \\ c^{ci} \end{bmatrix} = \begin{bmatrix} r_0^{ci} \\ c_0^{ci} \end{bmatrix} + \begin{bmatrix} s_u^{ci} & 0 \\ 0 & s_v^{ci} \end{bmatrix} \begin{bmatrix} u^{ci} \\ v^{ci} \end{bmatrix} \qquad (2)$$

being $[\, r_0^{ci} \quad c_0^{ci} \,]^{\mathrm{T}}$ the coordinates of the point O'_{ci} whereas s_u^{ci} and s_v^{ci} are the row and column scaling factor, respectively, for the i-th camera.

Consider an object frame O_o–$x_o y_o z_o$ attached to target object. The position and orientation of the object frame with respect to the base frame can be expressed in terms of the coordinate vector of the origin $o_o = [\, x_o \quad y_o \quad z_o \,]^{\mathrm{T}}$ and of the rotation matrix $R_o(\phi_o)$, where the components of $\phi_o = [\, \varphi_o \quad \vartheta_o \quad \psi_o \,]^{\mathrm{T}}$ are the Roll, Pitch and Yaw angles. The components of the vectors o_o and ϕ_o are the six unknown quantities to be estimated.

Consider m feature points of the object. The coordinate vector p_j^{ci} of the feature point P_j ($i = 1, \ldots, n, j = 1, \ldots, m$) can be expressed in the i-th camera frame as

$$p_j^{ci} = R_{ci}^{\mathrm{T}}(o_o - o_{ci} + R_o(\phi_o)p_j^o), \qquad (3)$$

where o_{ci} and R_{ci} are, respectively, the position vector and the rotation matrix of the i-th camera frame referred to the base frame, p_j^o is the coordinate vector of P_j expressed in the object frame. Notice that p_j^o is a constant vector that is assumed to be known, since it can be computed from a CAD model of the object or via a suitable calibration procedure. Moreover, the quantities o_{ci} and R_{ci} are constant, because each camera is assumed to be fixed to the workspace, and can be computed through a suitable calibration procedure [26].

By folding the $3m$ equations (3) into the perspective transformation (1) of the n cameras and into Eq. (2), a system of $2mn$ nonlinear equations is achieved. The equations depend on the measurements of the m feature points in the image plane of each cameras, while the six components of the vectors $\boldsymbol{o}_o$ and $\boldsymbol{\phi}_o$ are the unknown variables. To solve these equations at least six independent equations are required.

The computation of the solution is nontrivial and for visual servoing applications it has to be repeated at a high sampling rate. The recursive Kalman filter provides a computationally tractable solution, which can also incorporate redundant measurement information.

3   Kalman Filtering

In order to estimate the pose of the object, a discrete time state space dynamic model has to be considered, describing the object motion. The state vector of the dynamic model is chosen as the (12×1) vector

$$\boldsymbol{w} = [x_o \;\; \dot{x}_o \;\; y_o \;\; \dot{y}_o \;\; z_o \;\; \dot{z}_o \;\; \varphi_o \;\; \dot{\varphi}_o \;\; \vartheta_o \;\; \dot{\vartheta}_o \;\; \psi_o \;\; \dot{\psi}_o]^{\mathrm{T}}. \tag{4}$$

For simplicity, the object velocity is assumed to be constant over one sample period T. This approximation is reasonable in the hypothesis that T is sufficiently small. The corresponding dynamic modelling error can be considered as an input disturbance $\boldsymbol{\gamma}$ described by zero mean Gaussian noise with covariance given by the (12×12) matrix $\boldsymbol{Q}$. The discrete time dynamic model can be written as

$$\boldsymbol{w}_k = \boldsymbol{A}\boldsymbol{w}_{k-1} + \boldsymbol{\gamma}_k \tag{5}$$

where $\boldsymbol{A}$ is a (12×12) block diagonal matrix of the form

$$\boldsymbol{A} = \mathrm{diag}\left\{ \begin{bmatrix} 1 & T \\ 0 & 1 \end{bmatrix}, \cdots, \begin{bmatrix} 1 & T \\ 0 & 1 \end{bmatrix} \right\}.$$

The output of the Kalman filter, for each camera, is the vector of the *normalized* coordinates of the m feature points in the image plane of the camera

$$\boldsymbol{\zeta}_k = \begin{bmatrix} \dfrac{u_1^{c1}}{f_e^{c1}} & \dfrac{v_1^{c1}}{f_e^{c1}} & \cdots & \dfrac{u_1^{cn}}{f_e^{cn}} & \dfrac{v_1^{cn}}{f_e^{cn}} & \cdots & \dfrac{u_m^{c1}}{f_e^{c1}} & \dfrac{v_m^{c1}}{f_e^{c1}} & \cdots & \dfrac{u_m^{cn}}{f_e^{cn}} & \dfrac{v_m^{cn}}{f_e^{cn}} \end{bmatrix}_k^{\mathrm{T}}. \tag{6}$$

In view of (1), the corresponding output model can be written in the form

$$\boldsymbol{\zeta}_k = \boldsymbol{g}(\boldsymbol{w}_k) + \boldsymbol{\nu}_k \tag{7}$$

where $\boldsymbol{\nu}_k$ is the measurement noise, which is assumed to be zero mean Gaussian noise with covariance given by the $(2m \times 2m)$ matrix $\boldsymbol{R}$, and the function $\boldsymbol{g}(\boldsymbol{w}_k)$ is

$$\boldsymbol{g}(\boldsymbol{w}_k) = \begin{bmatrix} \dfrac{x_1^{c1}}{z_1^{c1}} & \dfrac{y_1^{c1}}{z_1^{c1}} & \cdots & \dfrac{x_1^{cn}}{z_1^{cn}} & \dfrac{y_1^{cn}}{z_1^{cn}} & \cdots & \dfrac{x_m^{c1}}{z_m^{c1}} & \dfrac{y_m^{c1}}{z_m^{c1}} & \cdots & \dfrac{x_m^{cn}}{z_m^{cn}} & \dfrac{y_m^{cn}}{z_m^{cn}} \end{bmatrix}_k^{\mathrm{T}}. \tag{8}$$

The coordinates of the feature points p_j^{ci} in (8) are computed from the state vector w_k via (3). Matrix R can be evaluated during the camera calibration procedure or by means of specific experiments.

Since the output model is nonlinear in the system state, it is required to linearize the output equations about the current state estimate at each sample time. This leads to the so-called Extended Kalman Filter (EKF).

The first step of the EKF algorithm provides an optimal estimate of the state at the next sample time according to the recursive equations

$$\hat{w}_{k,k-1} = A\hat{w}_{k-1,k-1} \tag{9}$$

$$P_{k,k-1} = AP_{k-1,k-1}A^{\mathrm{T}} + Q_{k-1}, \tag{10}$$

where $P_{k,k-1}$ is the (12×12) covariance matrix of the estimate state error. The second step improves the previous estimate by using the input measurements according to the equations

$$\hat{w}_{k,k} = \hat{w}_{k,k-1} + K_k(\zeta_k - g(\hat{w}_{k,k-1})) \tag{11}$$

$$P_{k,k} = P_{k,k-1} - K_k C_k P_{k,k-1}, \tag{12}$$

where K_k is the $(12 \times 2m)$ Kalman matrix gain

$$K_k = P_{k,k-1}C_k^{\mathrm{T}}(R_k + C_k P_{k,k-1}C_k^{\mathrm{T}})^{-1}, \tag{13}$$

being C_k the $(2m \times 12)$ Jacobian matrix of the output function

$$C_k = \left.\frac{\partial g(w)}{\partial w}\right|_{w=\hat{w}_{k,k-1}}. \tag{14}$$

The analytic expression of C_k can be found in the Appendix.

4   BSP Tree Geometric Modelling

The accuracy of the estimate provided by the Kalman filter depends on the number of the available feature points. Inclusion of extra points will improve the estimation accuracy but will increase the computational cost. It has been shown that a number of five or six feature points, if properly chosen, may represent a good trade-off [25]. Selection algorithms have been developed to find the optimal feature points [12]. In order to increase the efficiency of the selection algorithms, it is advisable to perform a pre-selection of the points that are visible to the camera at a given sample time. The pre-selection technique proposed in this chapter is based on Binary Space Partitioning (BSP) trees.

A BSP tree is a data structure representing a recursive and hierarchical partition of a n-dimensional space into convex subspaces. It can be effectively adopted to represent the 3D CAD geometry of an object [22].

In order to build the tree, each object has to be modelled as a set of planar *polygons*; this means that the curved surfaces have to be approximated. Each polygon

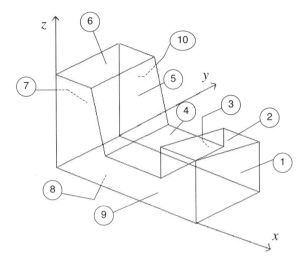

Fig. 2. Object and corresponding polygons.

is characterized by a set of *feature points* (the vertices of the polygon) and by the vector normal to the plane leaving from the object. For each node of the tree, a *partition plane*, characterized by its normal vector and a point, is chosen according to a specific criterion; the node is defined as the set containing the partition plane and all the polygons lying on it. The choice of the partition planes depends on how the tree will be used. For the purpose of removing the hidden surfaces, it is necessary to choose the partition planes in the set of the planes containing the polygons corresponding to the object surfaces.

The first node of the tree can be arbitrarily chosen. Different choices determine different trees. For the application considered here, the structure of the tree is not important because the visit algorithm must consider all the nodes.

Each node is the root of two subtrees: the *front* subtree corresponding to the subset of all the polygons lying entirely on the front side of the partition plane (i.e. the side corresponding to the half-space containing the normal vector), and the *back* subtree corresponding to the subset of all the polygons lying entirely on the back side of the partition plane.

The construction procedure can be applied recursively to the two subsets by choosing, for each node, a new partition plane among those corresponding to the polygons contained in that subtree.

If a polygon intersects the partition plane, it can be split into two or more pieces and the resulting parts are added to the corresponding subsets.

The construction ends when all the polygons and their parts are placed in a node of the tree.

As an example, consider the object represented in Fig. 2, which contains ten polygons. A possible BSP tree representation of the object is reported in Fig. 3, which

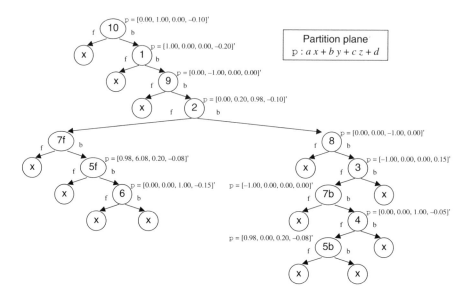

Fig. 3. BSP tree of the object.

has been obtained considering as root node the partition plane containing polygon number 10. A partition plane is represented by the vector $\pi = \begin{bmatrix} a & b & c & d \end{bmatrix}^T$ of the coefficients of the equation of the plane with respect to a base reference frame

$$ax + by + cz + d = 0,$$

where $n = \begin{bmatrix} a & b & c \end{bmatrix}^T$ is the unit vector normal to the plane. The root of the tree contains the polygon number 10; the front subtree is empty while the back subtree contains all the remaining polygons. The partition plane of the back subtree contains the polygon number 1; the front subtree is empty while the back subtree contains the polygons from number 2 to number 9. The construction ends when all the polygons are added to the nodes of the tree. Remarkably, the partition plane containing the polygon number 2 cuts polygons number 5 and 7 (notice that polygons number 9 and 10, which also intersect the partition plane, were already added to previous nodes of the tree), and thus they have been split into two pieces each (see polygons number 5f, 5b, 7f, 7b in Fig. 4).

In most cases, however, it is possible to choose the partition planes so that splitting of polygons is avoided. In this way the construction process of the tree and the visit algorithm are faster. This solution has to be preferred when the BSP tree must be built on line [18].

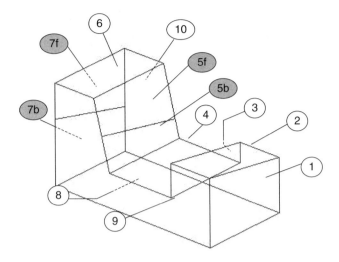

Fig. 4. Case of partition plane splitting polygons number 5 and number 7.

5   Features Selection

5.1   Pre-Selection Algorithm

Once a BSP tree representation of an object is available, it is possible to select the feature points of the object that are visible from a given camera position and orientation, by implementing a suitable visit algorithm of the tree. The algorithm can be applied recursively to all the nodes of the tree, starting from the root node as showed in Fig. 5, by updating a current set of visible feature points as follows.

For the current node, classify the camera position with respect to the current partition plane: **Front** side, **Back** side, **On** the plane. Hence:

- **Front:** Visit the back subtree; process the node; visit the front subtree.
- **Back:** Visit the front subtree; process the node; visit the back subtree.
- **On:** Visit the front subtree; visit the back subtree.

When the algorithm processes a node, the current set of projections of the visible feature points on the image plane is updated by adding all the projections of the feature points of the polygons of the current node and eliminating all the projections of the feature points that are hidden by the projections of the polygons of the current node.

If a polygon is hidden from the camera (i.e., the angle between the normal vector to the polygon and the camera z-axis is not in the interval $]-\pi/2, \pi/2[$ or the polygon is behind the camera), the corresponding feature points are not added to the set.

At the end of the visit, the current set will contain all the projections of the feature points visible from the camera, while all the hidden feature points will be discarded.

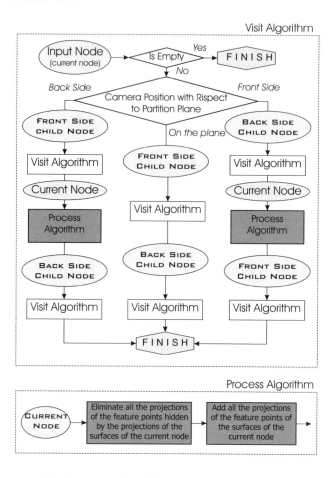

Fig. 5. Recursive visit algorithm of the BSP tree for the selection of visible feature points.

Notice that the visit algorithm updates the set by ordering the polygons with respect to the camera from the background to the foreground.

With reference to the BSP tree of Fig. 3, assuming that the camera is placed as the observer of the image in Fig. 4, the sequence of the processed nodes is: 10, 8, 7b, 4, 5b, 3, 2, 7f, 6, 5f, 9, 1, where the polygons number 10, 8, 7b, 3, 7f turn out to be hidden at the same time and will not be processed.

The technique described above can be suitably exploited to set up a real-time pre-selection algorithm of the feature points on the camera image plane, using the prediction of the estimated pose of the target object provided by the Kalman filter.

5.2   Selection Algorithm

The pre-selection technique recognizes all the feature points that are visible from a camera view point. However, this does not ensure that all the visible points are

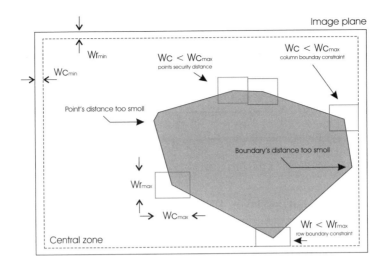

Fig. 6. Examples of significant situations during windowing test

"well" localizable, i.e., their positions can be effectively measured with a given accuracy. For instance, some points could be out of the field of view of the camera, or they could be too close to each other to guarantee absence of ambiguity in the localization. Moreover, the number of the well localizable feature points may be larger than the *optimal* number of points ensuring the best pose estimation accuracy.

In the following, a windowing test is adopted to select the projections of the feature points that can be well localized. Then, a selection algorithm is used to choose an optimal subset of points to be considered for feature extraction.

Windowing test The measurements of the coordinates of the projections of the feature points are obtained by considering suitable rectangular windows of the image plane to be grabbed and processed. Each window must contain one feature point. The windows are centered on the positions of the feature points on the image plane so as predicted by the Kalman filter. Their semi-dimensions are dynamically chosen in the interval $[Wr_{min}, Wr_{max}]$ for the base (the side parallel to the row's direction) and in the interval $[Wc_{min}, Wc_{max}]$ for the height (the side parallel to the column's direction). The minimum values are set so as to achieve a prescribed accuracy and robustness in the feature extraction, while the maximum values are set on the basis of the available memory and processing time.

A windowing test can be set up to select all the projections of the feature points that can be "well" localized.

First, all the points that are out of the field of view of the camera, or too close to the boundaries of the image plane, are discarded. This is achieved by eliminating all the points whose projections, so as predicted by the Kalman filter, are out of a central window of the image plane. The central window is obtained by reducing the height (base) of the whole image plane of the quantity Wr_{min} (Wc_{min}) from each side, as shown in Fig. 6.

Then, all the feature points that are too close to each other are discarded. This happens when the estimated distance between the projections of two or more points is lower than $S_f \cdot Wr_{min}$ ($S_f \cdot Wc_{min}$) along the row's (column's) direction; $S_f > 1$ is a suitable security factor.

All the remaining points are "well" localizable; the effective dimensions of the corresponding windows are dynamically adapted to the maximum allowable semi-dimension, so as to guarantee an assigned security distance from the other points and from the boundaries of the image plane (see Fig. 6).

Optimal feature points selection The number of feature points after pre-selection and windowing test is typically too high with respect to the minimum number sufficient to achieve the best Kalman filter precision. It has been demonstrated that an optimal set of five or six feature points guarantees about the same precision as that of the case when an higher number of feature points is considered [27,25].

The optimality of a set Γ of feature points is valued through the composition of suitably selected quality indexes into an optimal cost function. The quality indexes must be able to provide accuracy, robustness and to minimize the oscillations in the pose estimation variables. To achieve this goal it is necessary to ensure an optimal spatial distribution of the projections of the feature points on the image plan and to avoid chattering events between different optimal subsets of feature points chosen during the object motion. Moreover, in order to exploit the potentialities of a multi-camera system, it is important to achieve an optimal distribution of the feature points among the different cameras.

Without loss of generality, the case of two identical cameras is considered.

A first quality index is the measure of spatial distribution of the predicted projections on the image planes of a subset of q_i selected points for the i-th camera, $i = 1, 2$:

$$Q_{si} = \frac{1}{q_i} \sum_{k=1}^{q_i} \min_{\substack{j \in \{1, \dots, q_i\} \\ j \neq k}} \left\| \boldsymbol{p}_j - \boldsymbol{p}_k \right\|.$$

Notice that $q = q_1 + q_2$ is chosen between 6 and 8 to handle fault cases.

A second quality index is the measure of angular distribution of the predicted projections on the image planes of a subset of q_i selected points for the i-th camera, $i = 1, 2$:

$$Q_{ai} = 1 - \sum_{k=1}^{q_i} \left| \frac{\alpha_k}{2\pi} - \frac{1}{q_i} \right|$$

where α_k is the angle between the vector $\boldsymbol{p}_{k+1} - \boldsymbol{p}_{Ci}$ and the vector $\boldsymbol{p}_k - \boldsymbol{p}_{Ci}$, being $\boldsymbol{p}_{Ci}$ the central gravity point of the whole subset of feature points, and the q_i points of the subset are considered in a counter-clockwise ordered sequence with respect to $\boldsymbol{p}_{Ci}$, with $\boldsymbol{p}_{q_i+1} = \boldsymbol{p}_1$.

In order to avoid chattering phenomena, the following quality index, which introduces hysteresis effects on the change of the optimal combination of points, is considered for the i-th camera, $i = 1, 2$:

$$Q_h = \begin{cases} 1 + \epsilon & \text{if } \Gamma = \Gamma_{opt} \\ 1 & \text{otherwise} \end{cases}$$

where ϵ is a positive constant and Γ_{opt} is the optimal set of feature points at the previous sample time.

In order to distribute the points among the two cameras, the following indexes are considered:

$$Q_e = 1 + \frac{2}{q}\left(\frac{2}{q} - 1\right)\left|q_1 - \frac{q}{2}\right|$$

$$Q_d = \frac{q_1/d_1 + q_2/d_2}{q/\min\{d_1, d_2\}}$$

where q_i is the number of points assigned to the i-th camera, and d_i is the distance of the i-th camera form the object, $i = 1, 2$. The first index ensures an equal distribution of points among the cameras. The second index takes into account the distance of the cameras from the object, and thus allows managing different resolution zones of different cameras.

The proposed quality indexes represent only some of the possible choices, but guarantee satisfactory performance when used with the pre-selection method and the windowing test presented above, for the case of two fixed cameras. Other examples of quality indexes have been proposed [12], and some of them can be added to the indexes adopted here.

The cost function is chosen as

$$Q = Q_h \frac{Q_e Q_d}{q}\left(q_1 Q_{s1} Q_{a1} + q_2 Q_{s2} Q_{a2}\right)$$

and must be evaluated for all the possible combinations of the visible points on q positions. In order to determine the optimal set at each sample time, the initial optimal combination of points is first evaluated off line. Then, only the combinations that modify at most one point for camera with respect to the current optimal combination are tested on line, thus achieving a considerable reduction of processing time.

It should be pointed out that, in some cases, the number of points resulting at the end of the pre-selection step may bee too high to perform the optimal selection in a reasonable time. In such a cases, a computational cheaper solution, based on the optimal set at the previous time-step, can be adopted to find a sub-optimal set. For sufficiently small sampling time, the sub-optimal solution is very close or coincides with the optimal one.

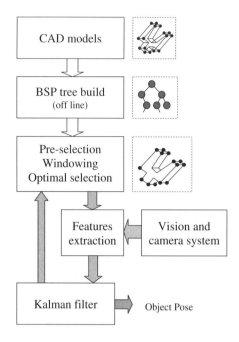

Fig. 7. Functional chart of the estimation procedure.

6   Estimation Procedure

A functional chart of the estimation procedure is reported in Fig. 7. It is assumed that a BSP tree representation of the object is built off-line from the CAD model. A Kalman filter is used to estimate the corresponding pose with respect to the base frame at the next sample time. The feature points selection and windows placing operation can be detailed as follows.

- **Step 1:** The visit algorithm described in the previous section is applied to the BSP tree of the object to find the set of all the feature points that are visible from the camera.
- **Step 2:** The resulting set of visible points is input to the algorithm for the selection of the optimal feature points.
- **Step 3:** The location of the optimal feature points in the image plane at the next sample time is computed on the basis of the object pose estimation provided by the Kalman filter.
- **Step 4:** A dynamic windowing algorithm is executed to select the parts of the image plane to be input to the feature extraction algorithm.

At this point, all the image windows of the optimal selected points are elaborated using a feature extraction algorithm. The computed coordinates of the points in the image plane are input to the Kalman filter which provides the estimate of the actual

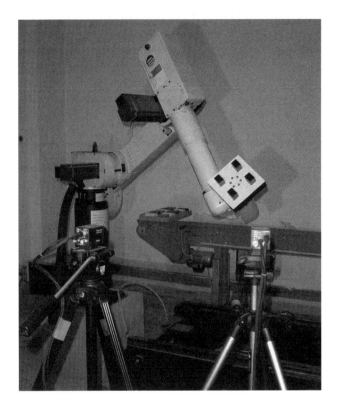

Fig. 8. Robot COMAU SMART3-S and SONY 8500CE cameras.

object pose and the predicted pose at the next sample time used by the pre-selection algorithm.

Notice that the procedure described above can be extended to the case of multiple objects moving among obstacles of known geometry [18]; if the obstacles are moving with respect to the base frame, the corresponding motion variables can be estimated using Kalman filters.

7   Experiments

7.1   Experimental Set-Up

The experimental set-up is composed by a PC with Pentium IV 1.7GHz processor equipped with two MATROX Genesis boards, two SONY 8500CE B/W cameras, and a COMAU SMART3-S robot (see Fig 8). The MATROX boards are used as frame grabber and for a partial image processing (e.g., windows extraction from the image). The PC host is also used to realize the whole BSP structures management, the pre-selection algorithm, windows processing, the selection algorithm and the

Kalman filtering. Some steps of image processing have been parallelized on the MATROX boards and on the PC, so as to reduce computational time. The robot is used to move an object in the visual space of the camera; thus the object position and orientation with respect to the base frame of the robot can be computed from joint position measurements via the direct kinematic equation. In order to test the accuracy of the estimation provided by the Kalman filter, the cameras were calibrated with respect to the base frame of the robot using the calibration procedure presented in [26], where the robot is exploited to place a calibration pattern in some known pose of the visible space of the cameras. The cameras resolution is 576×763 pixels and the nominal focal length of the lenses is 16 mm, while the calibration parameters for the two cameras are shown in Table 1. Notice that the parameters resulting from the calibration procedure are slightly different for the two cameras, although their nominal values are equal.

Table 1. Calibration parameters resulting from the calibration procedure.

Camera # 1
$r_0 = 187.96$
$c_0 = 318.20$
$f_u = -1955.84$
$f_v = 1953.41$
$\boldsymbol{o}_{c1} = [\, 1.2244 \ -1.7437 \ 0.8540 \,]^{\mathrm{T}} \ \mathrm{m}$
$\phi_{c1} = [\, -90.234° \ -1.880° \ 88.511° \,]^{\mathrm{T}}$
$\boldsymbol{d} = [\, 0.018 \ -0.019 \ -0.024 \ 0.012 \ 0.194 \,]^{\mathrm{T}}$

Camera # 2
$r_0 = 263.76$
$c_0 = 369.64$
$f_u = -1966.35$
$f_v = 1958.10$
$\boldsymbol{o}_{c2} = [\, 1.6149 \ -1.6565 \ 0.8623 \,]^{\mathrm{T}} \ \mathrm{m}$
$\phi_{c2} = [\, -92.188° \ -18.032° \ -89.111° \,]^{\mathrm{T}}$
$\boldsymbol{d} = [\, 0.009 \ 0.0048 \ -0.0027 \ -0.0096 \ 0.1393 \,]^{\mathrm{T}}$

Vector ϕ_{ci} contains the Roll, Pitch and Yaw angles of the i-th camera frame with respect to the base frame corresponding to the matrix $\boldsymbol{R}_{ci}$, while the vector $\boldsymbol{d} = [\, g_1 \ g_2 \ g_3 \ g_4 \ d_1 \,]^T$ contains the parameters used for compensating the distortion effects due to the imperfections of the lens profile and the alignment error of the optical system, as described in [26]. The estimated value of the residual mean triangulation error for the stereo camera system is 1.53 mm. The sampling time used for estimation is limited by the camera frame rate, which is about 26 fps. No particular illumination equipment has been used to test the robustness of the setup in the case of noisy visual measurements.

All the algorithms for BSP structure management, image processing and pose estimation have been implemented in ANSI C. The image features are the corners

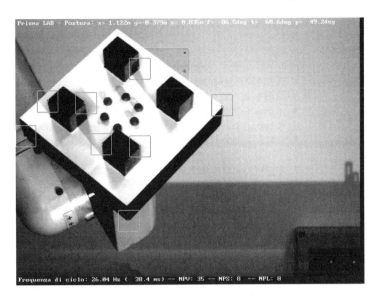

Fig. 9. Image seen by the camera with the windows selected for feature extraction. A point close to the center of each window marks the measured position of the corresponding feature point.

of the object, which can be extracted with high robustness in various environmental conditions. The feature extraction algorithm is based on Canny's method for edge detection [3] and on a simple custom implementation of a corner detector. In particular, to locate the position of a corner in a small window, all the straight segments are searched first, using an LSQ interpolator algorithm; then all the intersection points of these segments into the window are evaluated. The intersection points closer than a given threshold are considered as a unique average corner, due to the image noise. All the corners that are at a distance from the center of the window (which corresponds to the position of the corner so as predicted by the Kalman filter) greater than a maximum distance, are considered as fault measurements and are discarded. The maximum distance corresponds to the variance of the distance between the measured corner positions and those predicted by the Kalman filter.

The object used in the experiment is shown in Fig. 9, so as seen from the camera during the motion, as well as in Fig. 8, where the whole experimental setup is presented. The coordinates of the 40 vertices of the object, used as feature points, are reported in Table 2.

7.2   Experimental Results Using One Camera

Two different experiments have been realized for this case study. The first experiment reflects a favorable situation where the object moves in the visible space of the camera and most of the feature points that are visible at the initial time remain visible during

Table 2. Feature points coordinates with respect to the object frame, expressed in meters.

#	x^o	y^o	z^o	#	x^o	y^o	z^o
0	0.100	0.100	0.000	20	0.070	-0.039	0.092
1	0.100	-0.100	0.000	21	0.070	-0.070	0.092
2	-0.100	-0.100	0.000	22	0.029	-0.070	0.092
3	-0.100	0.100	0.000	23	0.029	-0.039	0.092
4	0.100	0.100	0.051	24	-0.029	-0.038	0.051
5	0.100	-0.100	0.051	25	-0.029	-0.069	0.051
6	-0.100	-0.100	0.051	26	-0.070	-0.070	0.051
7	-0.100	0.100	0.051	27	-0.070	-0.039	0.051
8	0.070	0.069	0.051	28	-0.029	-0.038	0.092
9	0.070	0.038	0.051	29	-0.029	-0.069	0.092
10	0.029	0.038	0.051	30	-0.070	-0.070	0.092
11	0.029	0.069	0.051	31	-0.070	-0.039	0.092
12	0.070	0.069	0.092	32	-0.028	0.069	0.051
13	0.070	0.038	0.092	33	-0.028	0.038	0.051
14	0.029	0.038	0.092	34	-0.069	0.039	0.051
15	0.029	0.069	0.092	35	-0.069	0.069	0.051
16	0.070	-0.039	0.051	36	-0.028	0.069	0.092
17	0.070	-0.070	0.051	37	-0.028	0.038	0.092
18	0.029	-0.070	0.051	38	-0.069	0.039	0.092
19	0.029	-0.039	0.051	39	-0.069	0.069	0.092

all the motion. The second experiment reflects an unfortunate situation where the set of the visible points is very variable, and a large part of the object goes out of the visible space of the camera during the motion.

The time history of the trajectory used for the first experiment is represented in Fig. 10. The maximum linear velocity is about 3 cm/s and the maximum angular velocity is about 3 deg/s.

The time history of the estimation errors is shown in Fig. 11. Noticeably, the accuracy of the system reaches the limit allowed by camera calibration, for all the components of the motion. As it was expected, the errors for some motion components are larger than others because only 2D information is available in a single camera system. In particular, the estimation accuracy is lower along z_c axis for the position, and about x_c and y_c axis for the orientation. Considering that in the experiment the z_c axis is almost aligned and opposed to the y axis of the base frame, the estimation errors are larger for the y component of the position, as well as the *roll* and *yaw* components of the orientation.

In Fig. 12 the output of the whole selection algorithm is reported. For each of the 40 feature points, two horizontal lines are considered: a point of the bottom line indicates that the feature point was classified as visible by the pre-selection algorithm at a particular sample time; a point of the top line indicates that the visible feature point was chosen by the selection algorithm. Notice that 8 feature points are selected at each sample time, in order to guarantee at least five or six measurements

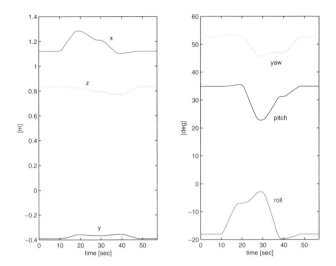

Fig. 10. Object trajectory with respect to the base frame used in the first experiment: position trajectory (*left*); orientation trajectory (*right*).

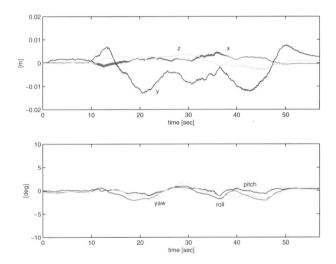

Fig. 11. Time history of the estimation errors in the first experiment: position errors (*top*); orientation errors (*bottom*).

in the case of fault of the extraction algorithm for some of the points. Also, some feature points are hidden during all the motion, while point number1 is only visible over some time intervals. Finally, no chattering phenomena are present.

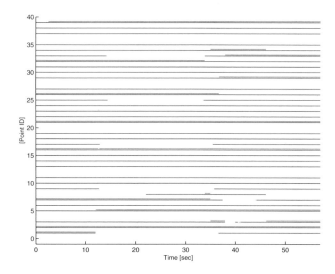

Fig. 12. Visible points and selected points in the first experiment. For each point, the bottom line indicates when it is visible, the top line indicates when it is selected for feature extraction.

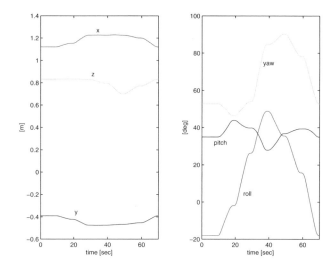

Fig. 13. Object trajectory with respect to the base frame used in the second experiment: position trajectory (*left*); orientation trajectory (*right*).

The time history of the trajectory used for the second experiment is represented in Fig. 13. The maximum linear velocity is about 2 cm/s and the maximum angular velocity is about 7 deg/s.

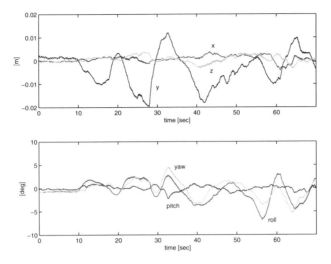

Fig. 14. Time history of the estimation errors in the second experiment: position errors (*top*); orientation errors (*bottom*).

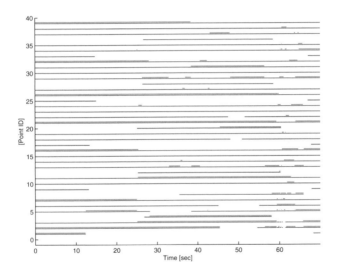

Fig. 15. Visible and selected points for the second experiment. For each point, the bottom line indicates when it is visible, the top line indicates when it is selected for feature extraction.

The time history of the estimation error is shown in Fig. 14. It can be observed that the error remains low but is greater than the estimation error of the previous experiment. This is due to the fact that from $t = 10$ s to $t = 60$ s the object moves so that it is partially out of the visible space of the camera; also, it rotates in such a way that a side remains almost parallel to the image plane. In this situation, just a few

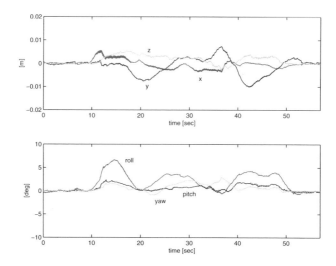

Fig. 16. Time history of the estimation errors in the case of two cameras: position errors (*top*); orientation errors (*bottom*).

feature points are visible; in addition, their projections on the image plane tend to be close or aligned so that the points that can be well localizable is further reduced and/or the spatial and angular distribution of the selected points is not optimal. This fact penalizes the estimation accuracy and explains how the magnitude of the estimation error components is one order greater than in the previous experiment, especially for the y component for the position error and the"roll" and "yaw" components of the orientation errors. The corresponding output of the pre-selection and selection algorithms are reported in Fig. 15. It should be pointed out that the pre-selection and selection algorithm are able to provide the optimal set of points independently from the operating condition, although slight chattering phenomena appear in some situation where the elements in set of localizable points is rapidly changing.

7.3   Experimental Results Using Two Cameras

The trajectory used for the experiment in the case of two cameras is the same represented in Fig. 10. The time history of the estimation errors is shown in Fig. 16. Noticeably, the accuracy of the system reaches the limit allowed by cameras calibration, for all the components of the motion, when the object does not move (about $5 \cdot 10^{-3}$ m for the position and about 1 deg for the orientation); during the motion the tracking errors grow but remain limited. As it was expected, the errors for the motion components are of the same order of magnitude, thanks to the use of a stereo camera system.

In Fig. 17 the output of the whole selection algorithm, for the two cameras, is reported. For each of the 40 feature points, two horizontal lines are considered: a point of the bottom line indicates that the feature point was classified as visible by the

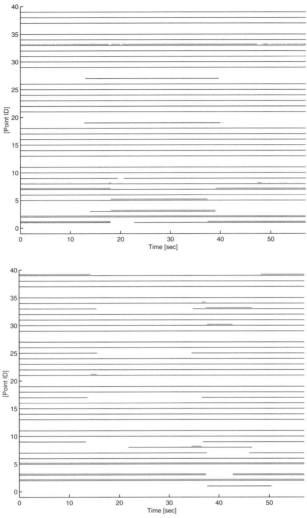

Fig. 17. Visible and selected points for Camera 1 (*top*) and Camera 2 (*bottom*), in the case of two cameras. For each point, the bottom line indicates when it is visible, the top line indicates when it is selected for feature extraction.

pre-selection algorithm at a particular sample time; a point of the top line indicates that the visible feature point was chosen by the selection algorithm. Notice that 8 feature points are selected at each sample time in order to guarantee at least five or six measurements in the case of fault of the extraction algorithm for some of the points. Remarkably, 4 feature points for camera are chosen at each sampling time, coherently with the almost symmetric disposition of the cameras with respect to the object.

8   Conclusion

The problem of real-time estimation of the pose (position and orientation) of a moving object from visual measurements has been considered in this chapter. A computationally efficient selection procedure has been presented, that allows evaluating the optimal set of feature points of the object to be used for image feature extraction and pose estimation. The procedure can be applied to polyhedral objects and is based on the representation of 3D objects by means of Binary Space Partitioning trees. The estimation technique fully exploits the noise rejection and the prediction capabilities of the EKF. Experimental results have been reported, which confirm the computational feasibility and the robustness of the presented visual tracking scheme for the case of two cameras.

The algorithm presented in this chapter may represent a good starting point to solve an important open issue for robotics applications: the visual tracking of objects in an unstructured and dynamic environment. A typical application may be the grasping of a moving object guided by a fixed visual system. In fact, for this scenario, the end effector may be considered as a second object of known pose. The proposed methodology may be used to develop a new strategy of automatic detection of the occlusions that happen during the grasp execution, which can be used to increase the task reliability. Similar problems may arise in cooperative robots applications.

Acknowledgement

This work has been co-funded by ASI.

References

1. K.S. Arun, T.S. Huang, and S.B. Blostein, "Least square fitting of two 3D point sets," *IEEE Trans. on Pattern Analysis and Machine Intelligence*, vol. 9, pp. 698–700, 1997.
2. J. Baeten and J. De Schutter, "Hybrid vision/force control at corners in planar robotic-contour following," *IEEE/ASME Trans. on Mechatronics*, vol. 7, pp. 143–151, 2002.
3. J. Canny, "A computational approach to edge detection," *IEEE Trans. on Pattern Analysis and Machine Intelligence*, vol. 8, pp. 679–698, 1986.
4. T.W. Drummond and R. Cipolla, "Real-time tracking of complex structures with on-line camera calibration," *British Machine Vision Conf.*, pp. 574–583, 1999.
5. B. Espiau, F. Chaumette, and P. Rives, "A new approach to visual servoing in robotics," *IEEE Trans. on Robotics and Automation*, vol. 8, pp. 313–326, 1992.
6. J.T. Feddema, C.S.G. Lee, and O.R. Mitchell, "Weighted selection of image features for resolved rate visual feedback control," *IEEE Trans. on Robotics and Automation*, vol. 7, pp. 31-47, 1991.
7. A. Fox and S. Hutchinson, "Exploiting visual constraints in the synthesis of uncertainty-tolerant motion plans," *IEEE Trans. on Robotics and Automation*, vol. 11, pp. 56–71, 1995.
8. P. C. Ho and W. Wang, "Occlusion culling using minimum occluder set and opacity map," *Proc. of 1999 IEEE Int. Conf. on Information Visualization*, pp. 292–300, 1999.

9. B.K.P. Horn, K.M. Hilden, and S. Negahdaripour, "Closed-form solution of absolute orientation using orthonormal matrices," *J. of Optical Society of America*, vol. A-5, pp. 1127-1135, 1998.

10. R. Horaud, F. Dornaika, and B. Espiau, "Visually guided object grasping," *IEEE Trans. on Robotics and Automation*, vol. 14, pp. 525–532, 1998.

11. S. Hutchinson, G.D. Hager, and P.I. Corke, "A tutorial on visual servo control," *IEEE Trans. on Robotics and Automation*, vol. 12, pp. 651–670, 1996.

12. F. Janabi-Sharifi and W.J. Wilson, "Automatic selection of image features for visual servoing," *IEEE Trans. on Robotics and Automation*, vol. 13, pp. 890-903, 1997.

13. F. Janabi-Sharifi and W.J. Wilson, "Automatic grasp planning for visual-vervo controlled robotic maniopulators," *IEEE Trans. on Systems, Man, and Cybernetics — Part B: Cybernetics*, vol. 28, pp. 693–711, 1998.

14. F. Keçeci and H.-H. Nagel, "Machine-vision-based estimation of pose and size parameters from a generic workpiece description," *Proc. of 2001 IEEE Int. Conf. on Robotics and Automation*, pp. 2159–2164, 2001.

15. S. Lee and Y. Kay, "An accurate estimation of 3-D position and orientation of a moving object for robot stereo vision: Kalman filter approach," *Proc. of 1990 IEEE Int. Conf. on Robotics and Automation*, pp. 414–419, 1990.

16. V. Lippiello, *Architetture, Algoritmi di Calibrazione e Tecniche di Stima dello Stato per un Sistema Asservito in Visione*, DIS, Univ. of Naples, Laurea Thesis, 2000.

17. V. Lippiello, B. Siciliano, and L. Villani, "Position and orientation estimation based on Kalman filtering of stereo images," *Proc. of 2001 IEEE Int. Conf. on Control Applications*, pp. 702–707, 2001.

18. V. Lippiello, B. Siciliano, and L. Villani, "Objects motion estimation via BSP tree modeling and Kalman filtering of stereo images," *Proc. of 2002 IEEE Int. Conf. on Robotics and Automation*, pp. 2968–2973, 2002.

19. E. Malis, F. Chaumette, and S. Boudet, "2 1/2 D visual servoing," *IEEE Trans. on Robotics and Automation*, vol. 15, pp. 234–246, 1999.

20. K. Nickels and S. Hutchinson, "Weighting observations: The use of kinematic models in object tracking," *Proc. of 1998 IEEE Int. Conf. on Robotics and Automation*, pp. 1677–1682, 1998.

21. J.N. Pan, Y.Q. Shi, and C.Q. Shu, "A Kalman filter in motion analysis from stereo image sequence," *Proc. of 1994 IEEE Int. Conf. on Image Processing*, pp. 63-67, 1994.

22. M. Paterson and F. Yao, "Efficient binary space partitions for hidden-surface removal and solid modeling," *Discrete and Computational Geometry*, vol. 5, pp. 485-503, 1990.

23. L. Sciavicco and B. Siciliano, *Modelling and Control of Robot Manipulators*, 2nd Ed., Springer Verlag, 2000.

24. K. Tarabanis, R. Y. Tsai, and A. Kaul, "Computing occlusion-free viewpoints," *IEEE Trans. on Pattern Analysis and Machine Intelligence*, vol. 18, pp. 279–292, 1996.

25. J. Wang and J.W. Wilson, "3D relative position and orientation estimation using Kalman filter for robot control," *Proc. of 1992 IEEE Int. Conf. on Robotics and Automation*, pp. 2638–2645, 1992.

26. J. Weng, P. Cohen, and M. Herniou, "Camera calibration with distortion models ad accuracy evaluation," *IEEE Trans. on Pattern Analysis and Machine Intelligence*, vol. 14, pp. 965–980, 1992.

27. J.W. Wilson, C.W. Hulls, and G. Bell, "Relative end-effector control using cartesian position based visual servoing," *IEEE Trans. on Robotics and Automation*, vol. 12, pp. 684–696, 1996.

28. J.S.-C. Yuan, "A general photogrammetric method for determining object position and orientation," *IEEE Trans. on Robotics and Automation*, vol. 5, pp. 129–142, 1999.

Appendix

The computation of the $(2mn \times 12)$ Jacobian matrix C_k in (14) gives

$$C_k = \left[\frac{\partial g}{\partial x_o} \ \ 0 \ \ \frac{\partial g}{\partial y_o} \ \ 0 \ \ \frac{\partial g}{\partial z_o} \ \ 0 \ \ \frac{\partial g}{\partial \varphi_o} \ \ 0 \ \ \frac{\partial g}{\partial \vartheta_o} \ \ 0 \ \ \frac{\partial g}{\partial \psi_o} \ \ 0 \right]_k \quad (15)$$

where 0 is a null $(2mn \times 1)$ vector corresponding to the partial derivatives of g with respect to the velocity variables, which are null because function g does not depend on the velocity.

Taking into account the expression of g in (8), the non-null elements of the Jacobian matrix (15) have the form:

$$\frac{\partial}{\partial \alpha}\left(\frac{x_j^c}{z_j^c}\right) = \left(\frac{\partial x_j^c}{\partial \alpha}z_j^c - x_j^c\frac{\partial z_j^c}{\partial \alpha}\right)(z_j^c)^{-2} \quad (16)$$

$$\frac{\partial}{\partial \alpha}\left(\frac{y_j^c}{z_j^c}\right) = \left(\frac{\partial y_j^c}{\partial \alpha}z_j^c - y_j^c\frac{\partial z_j^c}{\partial \alpha}\right)(z_j^c)^{-2} \quad (17)$$

where $\alpha = x_o, y_o, z_o, \varphi_o, \vartheta_o, \psi_o$, $i = 1,\ldots,n$, $j = 1,\ldots,m$.

The partial derivatives on the right-hand side of (16) and (17) can be computed as follows.

In view of (3), the partial derivatives with respect to the components of vector $o_o = [x_o \ \ y_o \ \ z_o]^T$ are the elements of the Jacobian matrix

$$\frac{\partial p_j^c}{\partial o_o} = R_c^T.$$

In order to express in compact form the partial derivatives with respect to the components of the vector $\phi_o = [\varphi_o \ \ \vartheta_o \ \ \psi_o]^T$, it is useful to consider the following equalities [23]

$$d R_o(\phi_o) = S(d\omega_o)R_o(\phi_o) = R_o(\phi_o)S(R_o^T(\phi_o)d\omega_o) \quad (18)$$

$$d\omega_o = T_o(\phi_o)d\phi_o \quad (19)$$

where $S(\cdot)$ is the skew-symmetric matrix operator, ω_o is the angular velocity of the object frame with respect to the base frame, and the matrices R_o and T_o, in the case of Roll, Pitch, Yaw angles, have the form

$$R_o(\phi_o) = \begin{bmatrix} c_{\varphi_o}c_{\vartheta_o} & c_{\varphi_o}s_{\vartheta_o}s_{\psi_o} - s_{\varphi_o}c_{\psi_o} & c_{\varphi_o}s_{\vartheta_o}c_{\psi_o} + s_{\varphi_o}s_{\psi_o} \\ s_{\varphi_o}c_{\vartheta_o} & s_{\varphi_o}s_{\vartheta_o}s_{\psi_o} + c_{\varphi_o}c_{\psi_o} & s_{\varphi_o}s_{\vartheta_o}c_{\psi_o} - c_{\varphi_o}s_{\psi_o} \\ -s_{\vartheta_o} & c_{\vartheta_o}s_{\psi_o} & c_{\vartheta_o}c_{\psi_o} \end{bmatrix}$$

$$T_o(\phi_o) = \begin{bmatrix} 0 & -s_{\varphi_o} & c_{\varphi_o}c_{\vartheta_o} \\ 0 & c_{\varphi_o} & s_{\varphi_o}c_{\vartheta_o} \\ 1 & 0 & -s_{\vartheta_o} \end{bmatrix},$$

with $c_\alpha = \cos\alpha$ and $s_\alpha = \sin(\alpha)$. By virtue of (18), (19), and the properties of the skew-symmetric matrix operator, the following chain of equalities holds

$$\mathrm{d}(\boldsymbol{R}_o(\boldsymbol{\phi}_o)\boldsymbol{p}_j^o) = \mathrm{d}(\boldsymbol{R}_o(\boldsymbol{\phi}_o))\boldsymbol{p}_j^o = \boldsymbol{R}_o(\boldsymbol{\phi}_o)\boldsymbol{S}(\boldsymbol{R}_o^T(\boldsymbol{\phi}_o)\boldsymbol{T}_o(\boldsymbol{\phi}_o)\mathrm{d}\boldsymbol{\phi}_o)\boldsymbol{p}_j^o$$
$$= \boldsymbol{R}_o(\boldsymbol{\phi}_o)\boldsymbol{S}^T(\boldsymbol{p}_j^o)\boldsymbol{R}_o^T(\boldsymbol{\phi}_o)\boldsymbol{T}_o(\boldsymbol{\phi}_o)\mathrm{d}\boldsymbol{\phi}_o$$
$$= \boldsymbol{S}^T(\boldsymbol{R}_o(\boldsymbol{\phi}_o)\boldsymbol{p}_j^o)\boldsymbol{T}_o(\boldsymbol{\phi}_o)\mathrm{d}\boldsymbol{\phi}_o,$$

hence

$$\frac{\partial \boldsymbol{R}_o(\boldsymbol{\phi}_o)}{\partial \boldsymbol{\phi}_o}\boldsymbol{p}_j^o = \boldsymbol{S}^T(\boldsymbol{R}_o(\boldsymbol{\phi}_o)\boldsymbol{p}_j^o)\boldsymbol{T}_o(\boldsymbol{\phi}_o). \tag{20}$$

At this point, by virtue of (3) and (20), the following equality holds

$$\frac{\partial \boldsymbol{p}_j^c}{\partial \boldsymbol{\phi}_o} = \boldsymbol{R}_c^T\frac{\partial \boldsymbol{R}_o(\boldsymbol{\phi}_o)}{\partial \boldsymbol{\phi}_o}\boldsymbol{p}_j^o = \boldsymbol{R}_c^T\boldsymbol{S}^T(\boldsymbol{R}_o(\boldsymbol{\phi}_o)\boldsymbol{p}_j^o)\boldsymbol{T}_o(\boldsymbol{\phi}_o).$$

RTLinux-Based Controller for the SuperMARIO Mobile Robot

Claudio Bellini[1], Stefano Panzieri[1], Federica Pascucci[2], and Giovanni Ulivi[1]

[1] Dipartimento di Informatica e Automazione
Università di Roma Tre
Via della Vasca Navale 79, 00146 Roma, Italy
<bellini,panzieri,ulivi>@dia.uniroma3.it
http://www.labrob.it

[2] Dipartimento di Informatica e Sistemistica
Università di Roma "La Sapienza"
Via Eudossiana 18, 00184 Roma, Italy
pascucci@dia.uniroma3.it

Abstract. In the last years a new way to implement Real Time control systems has been opened, in connection with the diffusion of the open-source operating system Linux. There are several proposals to force this system to become or, at least, to behave as a Real Time one. Some of them are open source as the original operating system. The purpose of this work is two-fold. First, to describe the mechanical structure and the electronics of the mobile robot SuperMARIO (Mobile Autonomous Robot for Indoor Operations): this unit was built in our laboratory about three years ago and the design was oriented to high precision trajectory tracking and high dynamic performance. Second, to detail the software architecture based on the RTLinux OS, including low level Real Time motor feedback, high level trajectory loops, and communications protocols that, through an IEEE 802.11 radio link, allow the interaction with remote computers as a part of our laboratory network.

1 Introduction

Often, in the field of mobile robotics, two different choices are at stakes: to buy a ready-made unit or to build a dedicated prototype. This is in particular true when the project includes the low level (motor) control. It is very difficult, indeed, to gain access to this level in commercial systems that seem to be more oriented to researches in the high level part of the control structure and in general do not provide access to the source code.

The same happened to our group. We already had two commercial units, from two different producers, but none of them is completely satisfactory from the control point of view. Even the sensor sampling time cannot be modified, not to say the motor controller parameters.

These considerations revamped an old project; namely, the SuperMARIO (Mobile Autonomous Robot for Indoor Operations) which was first developed, at the Robotics Laboratory of University of Rome "La Sapienza" [7], mainly as a high precision platform to test sophisticated control algorithms. *Super* was added to differentiate this robot from a previous one named MARIO (see [8]) with a less precise

B. Siciliano et al. (Eds.): Advances in Control of Articulated and Mobile Robots, STAR 10, pp. 153–169, 2004.
© Springer-Verlag Berlin Heidelberg 2004

mechanical structure. The availability of powerful processors and mainly our interest in testing new real-time operating systems, made the first SuperMARIO a good platform to start with.

Aim of this chapter is to describe the overall structure of the new SuperMARIO that shows substantial differences in both low-level motor control and software architecture with respect to its predecessor, so that other groups may gain information and understand about the pros and cons of undertaking such a work.

2   The New SuperMARIO Mobile Robot

2.1   Electro-Mechanical Structure

SuperMARIO is a two-wheel differentially driven robot. An aluminum chassis, two actuated wheels on the rear axle and a front castor compose its mechanical structure.

The chassis is 3 mm thick and measures $45 \times 32 \times 32$ cm. The chassis is composed of two compartments. The lower contains the two motors, the rear axle and the transmission elements, while the power supply system (i.e., two 12 V batteries and some power supplies) takes place in the upper compartment together with the power electronics. The front side of the chassis is equipped with an ISA backplane, in which a single board computer Intel 486 DX/4 100MHz, a wireless Ethernet device and the motor interface board are connected.

The actuated wheels have a radius $r = 9.5$ cm and were machined by a lathe for maximum accuracy. A stiff O-ring is used to prevent slippage and ensure a small contact surface with the ground. The wheels are actuated by two MCA permanent magnet d.c. servo motors. This kind of actuator presents a good power/dimension ratio with respect to the stepping motor and is easier to control than a brushless one. Unfortunately, it is affected by torque ripple at low speeds and needs a velocity transducer; so each motor is equipped with an incremental encoder with 200 pulse/turn. A syncroflex planetary gearbox with a reduction ratio 20:1 is used to reduce the velocity and improve the odometric measures. In order to eliminate the disturbances induced by reorientation of the castor, a spherical bearing is placed in front of the vehicle.

3   The Motor Interface

To fully control a d.c. motor by a computer, some functions are necessary: in the forward path we typically find a PWM modulator (that translates a value in a suitable two-level waveform with the desired short term mean value) and a power amplifier. On the feedback path we have a detector for the sign of the rotation and an up/down counter to measure the axle angle variation in a sampling period. These functions must be duplicated for both motors. Moreover, a connection is needed between these functions and the computer.

The cards available from the market are typically very complex and have capabilities far beyond those needed by our project, so their cost is rather high[1]. Therefore,

---

[1] This is the result of a small Internet research with the constraint of Linux compatibility.

we decided to design and build a card implementing just the above described functions.

An important factor contributing to this choice was the availability of field programmable devices that allow the implementation of complex digital networks. They make a large part of the testing and debugging phase almost as easy as that of a software routine. Indeed, the hardware part (whose errors can force to redesign the whole card) is reduced to some supporting logic gates to interface the device to the computer bus. In particular we decided to use an Altera FPGA (Field Programmable Gate Array) MAX7128SLC84-10 that contains 128 logic cells.

It can be on-board reprogrammed by a dedicated programmer connected to the serial port of the PC used to develop the project. The description of the circuit can be entered in a sort of high level language —we used VHDL, in which more independent "entities", i.e. subprojects, can be developed— that can be compiled and downloaded into the FPGA. A simulator is available to check the design before downloading the code.

The FPGA project is composed by three entities:

• encoders signal decoding,
• generation of modulated signals,
• ISA bus interfacing.

n

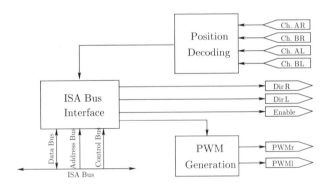

Fig. 1. Structural logic scheme.

In Fig. 1 a scheme of the three entities developed is reported. Inputs, outputs and blocks interconnections are shown. In particular, the ISA interface is connected with the ISA bus and the two blocks for position decoding and PWM generation. Moreover it provides the digital signals to enable the motor drivers (Enable) and to select the direction of motor rotation (Dir R and Dir L).

3.1   Position Decoding

Each encoder provides two channels (A and B). The two channels, at constant speed, produce two square waves with a phase delay of $\pi/2$. Their frequency is proportional to rotation speed and, if B channel has a time lag with respect to A channel the encoder has a clockwise rotation. On the contrary, if time lag is in A channel the encoder has a counterclockwise rotation. The logic implemented to determine the number of impulses and the direction of rotation in a robust way is shown in Tab. 1.

Table 1. Determination of rotation direction.

CH A	CH B	Direction
↑	0	CW
↑	1	CCW
↓	0	CCW
↓	1	CW
0	↑	CCW
1	↑	CW
0	↓	CW
1	↓	CCW

Although this function can be implemented in an asynchronous way, the FPGA has a synchronous behavior with an internal clock at 8 MHz. To cope with this limitation four D-type flip-flops, a chain of two for each channel, have been used to synchronize A and B signals and to make available their previous values. The implemented logic is reported in Fig. 2.

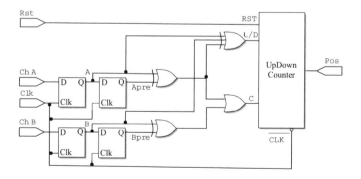

Fig. 2. Logic for a single encoder.

Using an XOR between X and Xpre (being X any channel), a pulse is obtained when a transition happens on a channel. An OR between the two XOR's gives a signal that has a pulse for each channel transition. This signal feeds the enable of an up/down counter whose clock is connected to the falling edge of the FPGA clock. We use the falling edge to give time all the FPGA transitions to be stabilized; this happens on the rising edges of the clock. Instead of using the logic function described in Tab. 1 to determine the rotation direction, we simply employed a three-input XOR, observing that it is sufficient to determine on which channel there has been the last transition and whether channels are on the same level or not after this event.

The decoding function is implemented two times for each encoder. Two additional 8 bit registers latch the values of the two counters at the same time; their values are then transferred to the CPU via the ISA bus.

Fig. 3. Single encoder entity.

3.2   Generation of PWM

The two DC motors are fed with a pulse wide modulation (PWM) signal. For each motor the microcontroller sends to the FPGA card the sign of the supply voltage and the duty cycle of the modulation. Duty cycle is given as an unsigned 8 bit representation.

As shown in Fig. 4, the modulated signals PWMr and PWMl can be obtained comparing the duty-cycle values with one linear ramp between 0 and 255. The output will be high when the ramp is greater that this number.

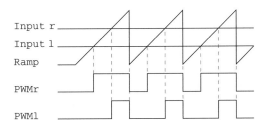

Fig. 4. PWM generation.

The clock of the counter is equal to 8 MHz resulting in a 31250 KHz modulation.

3.3   ISA Bus Interfacing

Communications among the cards of the control computer is obtained through ISA bus. Let us analyze shortly its protocol: an ISA bus is a cluster of three buses:

- an 8-bit data bus;
- a 20-bit address bus, where the first 12 are actually used with addresses ranging from 000 to FFF;
- a 3-bit control bus (IOR, IOW e AEN).

The motherboard has always full control on the address bus and on signals IOR and IOW (read and write). To read data, the CPU sets its address and force the IOR bit to zero: a three-state register is enabled and a value is written on the data bus. Viceversa, to write data, the CPU sets the address, writes the value on the data bus and forces the IOW bit to zero. Bit AEN, involved in DMA operations, is always zero. Each card on an ISA bus has a selectable base address (BA) that points to an internal register that can be read or written; other registers are at addresses BA+1, BA+2, etc. In our card, the BA is set by a DIP switch (SW1).

A simple discrete logic, shown in Fig. 5, performs the recognition of the base address and alerts the FPGA zeroing the Outnand signal; then the FPGA decodes the three less significative bits on the address bus to select a particular register. In Tab. 2 the role of each register is reported:

Table 2. Role of registers for the FPGA.

Address	Reading	Writing
BA + 0	left motor position	not used
BA + 1	right motor position	not used
BA + 2	not used	left motor PWM
BA + 3	not used	right motor PWM
BA + 4	not used	digital outputs
BA + 5	not used	reset

To read left motor position is, for instance, sufficient to read the value of register AB+0, and this can be easily done in C language with instruction `data=inportb(AB+0)`. In the same way, to set duty cycle for right motor instruction `outportb(BA+3,data)` can be used. The digital output register is used in two ways: first to send direction and enable bits to the motor drives, second to produce the latch signal that forces counters data to be synchronously memorized in registers BA+0 and BA+1. Afterwards, these registers can be read.

Finally, the less significant bit of register BA+5 is used to reset the FPGA.

After an extended simulation of the VHDL software, the FPGA has been programmed and the board has been built. An 8MHz oscillator provides the clock signal. In Fig. 6 the final board is shown.

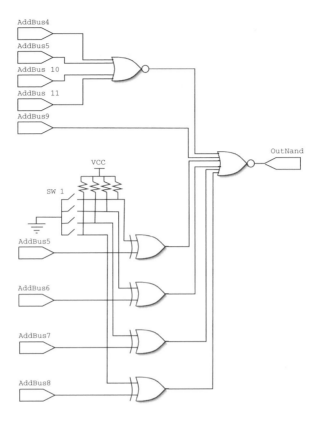

Fig. 5. Outnand generation.

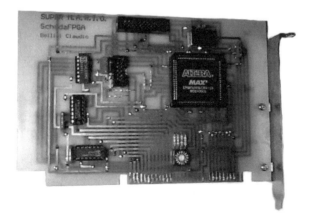

Fig. 6. FPGA board.

4   The Motor Control Algorithm

The mobile robot is driven by two small permanent magnet d.c. motors, each one coupled to a wheel by a planetary gearbox (see Tab. 3 for the main parameters). Designing a PI controller around the rotation speed is, in principle, an easy homework for a first-level course in Automatic Control. However, when all the nonlinearities are taken into account a more sophisticated algorithm is needed, in particular to obtain a smooth run at low speeds.

Table 3. Motor parameters.

Motor parameters		
K_m	$0,056$	$N\ m/A$
R	$2,1$	Ω
L	$2,6$	mH
D	$0,00045$	$N\ m\ s/rad$
J	$0,00015$	$N\ m$

The main nonlinearities affecting the system behavior and thus the controller design are the discretization introduced by the encoder and the dry friction in the gearbox.

The first can be modeled as shown in Fig. 7.

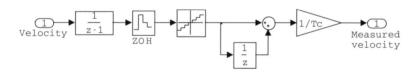

Fig. 7. Model of speed estimation.

As usual, the velocity is computed by the difference of two successive position measures. An adaptive scheme has been proposed in [3], that at low speeds changes its behavior to counting the clock pulses between two (or more) encoder pulses. It is however too complex for the FPGA implementation we chose.

The discretization acts on the angle measures (that are obtained by integration of the wheel speed) and its step is equal to 0.039mm. Clearly, the resolution of the speed measurements is proportional to the sampling time. A higher time gives a better resolution; however it also gives a worse dynamic behavior to the system. As a tradeoff, a 5ms sampling time has been chosen, which is far within the capabilities of the computer. With this choice, the velocity resolution turns out to be equal to 7.8 mm/s.

An accurate simulation of the dry friction at the motor shaft is a difficult task (e.g., [1]; here we used a simplified model that gave results sufficient for our design purposes. It is shown in Fig. 8 as a Simulink scheme.

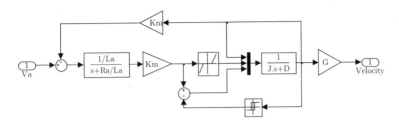

Fig. 8. Motor model including friction.

At very low speeds, the electromagnetic torque is fed into a deadzone, whose band represents the stiction phenomenon. The linear parts have unitary slope. Until the modulus of provided torque is lower than the stiction one, no mechanical torque is applied to the load. When the velocity is different from zero, the selector moves to the other input and a constant torque (with the proper sign) is subtracted from the electromagnetic one. The hysteresis has a width equal to that of the deadzone to guarantee continuity during the switching. The viscous part of the friction (that is very little compared to the dry part) is modeled as D in the mechanical load (where J represents the inertia).

The amplifier saturation and command discretization are also modeled in the simulation. Both phenomena have minor effects in the low speed behavior of the system.

Figure 9 shows the simulated speed obtained with the model including the nonlinearities and a simple Proportional-Integral controller when the desired speed is a step in $t = 0$ with an amplitude equal to 10 mm/s.

The initial lag is due to stiction, while the "noise" is actually the effect of the speed discretization. Note that the latter phenomenon bans the use of a derivative action in the controller.

To cope with the two described phenomena, two very simple, yet effective, modifications have been applied to the basic controller. Both are feedforward term, the first constant and not influenced (except for the sign) by the input value (Gain1), the second proportional to the input itself. The overall scheme is given in Fig. 10.

As for the dry friction, a very simple feedforward has been applied. Its amplitude is a little lower that the estimated stiction. This solution, in general, is not very robust vs. parameter variations and adaptive scheme could be devised. However in typical experimental conditions and rather repetitive environment temperature this sophistication is not needed. Also a small linear feedforward is used to improve the response speed of the system.

With this add-on, the response for the same input used for Fig. 10 becomes that referred in Fig. 11, which shows a satisfactory initial transient.

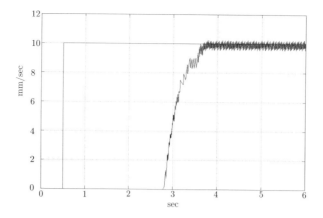

Fig. 9. Step response (10mm/s) with a simple controller.

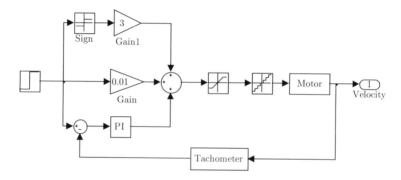

Fig. 10. Scheme with feedforward.

The quantization effects on the velocity measures have been treated as a noise. As a matter of fact, the "quantization noise" is uncorrelated with the original signal. Therefore, a second order filter has been designed with a natural frequency equal to 70 rad/s and a damping coefficient equal to 0.5.

With this last change, the response in the same test conditions is that shown in Fig. 12. It can be seen that the effects of the quantization noise are greatly reduced, but at the cost of an overshoot at the beginning of the transient. A different filter can reduce this effect but would also slow the response (bandwidth) of the system; so we decided to keep the chosen filter, as it gives a fast disturbance rejection, but we designed the outer level so that the reference speed is smooth, thus avoiding high overshoots [4] [6].

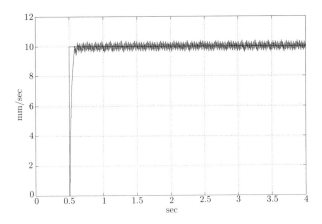

Fig. 11. Step response with added feedforward.

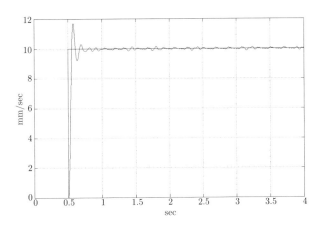

Fig. 12. Step response with feedforward regulator with digital filtering.

5   The RTLinux Architecture

Operating systems normally used for *office* applications are not suitable to implement control algorithms. Our interest has been concentrated on GNU/Linux operating system which, compared to Microsoft systems, has the great advantage of being completely Open source and being based on the Unix system. As all Unix systems, Linux scheduler is *preemptive*, that means that a process can always loose the processor utilization when another process has matured a greater priority. Most recent versions of Linux kernel [2] introduce the possibility to place side by side a static priority, definable from the user, and a dynamic priority, periodically calculated from the scheduler². All normal processes have 0 as static priority; therefore a process with a priority greater than zero will be favorite for the processor utilization. Kernel processes remain however excluded from the normal priority mechanism, they can always interrupt the other processes and temporary take the exclusive use of the processor, inhibiting the possibility to have a context change. A scheduling algorithm like the one described gives good results in the management of normal activities but is not suitable for real-time applications. To be able to guarantee sufficiently precise sampling times and to assure that the control related computations take a short time, it is necessary that a process is able to obtain the exclusive utilization of processor within a well-know time.

5.1   Real Time Linux

To overcome the limitations due to the use of the Unix scheduler, several techniques are evolving that make Unix a system suitable to execute hard real-time applications³. To improve the support to real-time applications, Linux, as many other Unix-based systems, conforms, in part, to POSIX.1b-1993 standard. This standard introduces a scheduler with user definable static priorities and the possibility to execute more than one thread in a single process. Usually, only one program counter is used to execute a block of instructions in a process; according to the POSIX standard it is possible to run more than one block or instruction side by side in the same process. Hence it is possible to design a cooperating threads architecture to optimize process resource handling.

Unfortunately there are still some unsolved problems, such as:

1. *not-preemptability* of kernel processes,
2. low clock resolution,
3. high wait time for IRQ response.

Various techniques, based on that standard, have been developed to solve these problems, permitting to execute hard real-time tasks in Unix-like systems. One of these solutions, that has the characteristic of being completely free and Open source, is called *Real Time Linux* [12,11,14,13,9,5]. The greater obstacle for the execution of real-time tasks is the first listed point; kernel processes use specific processor instructions (e.g., `cli` and `sti` for Intel family processors) to disable the interrupts. In Real Time Linux a software layer has been inserted between the request to disable interrupts and the effective call of `cli` and `sti`; this layer allows preventing the interrupt of selected tasks from other processes [2].

---

² In less recent versions, the scheduling algorithm, to optimize processor allocation, calculates only the priority of active processes with a regular period; in this case we speak of *dynamic priority*.

³ Two different Real Time applications can be defined: those that need more accurate sampling times are called *Hard Real Time* applications, those that, instead, do not need particularly stringent performances are called *Soft Real Time* applications[13].

Regarding points 2 and 3, it has been possible to obtain for the IRQ response a resolution of approximately $15\mu s$ in the worst case, taking advantage of the built-in timer on Intel 8354 chip, present on all IBM compatible PC.

Through these tools, RTLinux provides some APIs that permit building real-time applications with performances suitable for our application.

6   RTLinux Control Architecture and Communication Protocol

A typical control application in RTLinux environment is composed of a low level layer and a high level one.

The lower level layer is implemented in a kernel module where the Real Time threads run. Each thread consists in a set of instructions executed periodically; the maximum time spent in order to execute an iteration of this cycle represents the minimum sampling time definable for the corresponding control function. Among the several functions of the RTLinux APIs, there are some that allow regulating the iteration time with great precision, permitting the designer to choose the sampling time he/she prefers. It is important to remark that, during the wait time between two sampling intervals, the processor is free and it can, therefore, be used for other applications.

The high level part, instead, consists of a process, running in user space, that manages TCP socket connections with remote clients and sends commands, data and references to the low-level part. This application can handle different connections with some clients, each dedicated to one of a set of services with different complexity. As all the operating system processes (e.g., user applications) become active only when Real Time threads are in a wait state, the designer has to consider which percentage of time is used for these operations and which is available for other processes. Taking care of this percentage, it is possible to determine the complexity of high level applications.

Figure 13 shows an outline of the software architecture for a typical RTLinux application.

To allow communications between user processes and Real Time threads there are appropriate structures named RT-FIFOs. These are seen from the kernel level as queues where it is possible to read or write blocks of characters by the typical operations `rtf_get` and `rtf_put`. Since an RT-FIFO structure is one way, in order to obtain a bidirectional data flow it is necessary to instantiate two separate structures. At the user level these are seen as character devices (`/dev/rft*` where it is possible to read or write blocks of text by the standard library functions `write` and `read`. Viewing a couple of FIFOs as a single FIFO at user level is possible thanks to the `rtf_make_user_pair` command.

Using Open source software, all the system source code is available. In RTLinux it is possible to implement a scheduling algorithm, designed for a specific application, simply loading the appropriate kernel extension module [13].

In the released RTLinux 3.0 version a very simple priority preemptive scheduler is provided: a priority is *statically* assigned to every process, when more than one

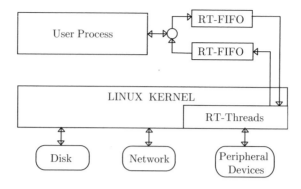

Fig. 13. Software architecture for an RTLinux application.

task is ready; the one with the greatest priority is executed. If a task with greater priority becomes ready it immediately interrupts the task in execution; moreover each task releases the CPU when the critical real-time block is terminated.

This scheduler supports periodic applications, and is possible to execute isolated tasks defining an interrupt handler. Linux is, for this scheduler, the Real Time process with lower priority; in this way the system is ready for other applications only when no Real Time thread is in execution.

6.1 TCP Connection Manager

To manage connections with external processes, we implemented a TCP socket server running in the user space. A parent process is always waiting for new connections on a dedicated port. When a client tries to connect to SuperMARIO, the server identifies the class of that client (e.g., "movement manager", "vision sensor" etc . . .). If a client of the same class is already connected the server closes the connection, otherwise it creates a child to manage the connection. The parent keeps a list of all the children created; in such a way it can kill all of them when the user decides to turn off the server. Each child receives commands, data and references as a structured message. The child elaborates the message and if necessary (it depends on the kind of message) it sends it to the kernel module without modifying it. To do that it uses the bidirectional RT-FIFO described before. When a child has to send a message to the kernel module it is important to verify that no other child is using the FIFO. To do that, there is a semaphore to indicate the state of the FIFO [10].

6.2 Threads Architecture

As reported above, our application is composed from a low-level and a high-level layer. In the latter the Connection Manager is implemented to provide an interface with external clients.

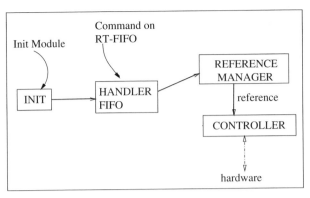

Fig. 14. Low-level architecture.

Figure 14 details the architecture of the kernel module. There are two Real Time threads running simultaneously two different tasks. One is the motor control algorithm (CONTROLLER), the other provides the reference values to the controllers (REFERENCE MANAGER), according to the kind of job specified by a suitable "Movement Manager Client". Obviously the CONTROLLER priority is higher than the REFERENCE MANAGER one. On the contrary, the Controller sampling time is generally shorter than that of the Manager.

The global architecture for the Motor Controller is obtained implementing the one previously described in Section 4. The sequence of operations described below is executed in each control cycle:

1. encoder reading,
2. output of voltage values computed in the previous cycle,
3. computation of the next control action.

6.3   The Communication Protocol

We decided to use the same communication protocol at any control level: on the client-server communications and on the server-kernel one. We used a structured message where the first field is an integer value that specifies the command name. The meaning of the other fields of the message depends on the value of the first field. First of all, when a child receives a message it decodes the first field to understand whether it has to do something or else it is only necessary to send the same message to the kernel module without modifying it. For instance, when the Movement Client sends the command "Stop Server", the receiving child forwards immediately the same message to the kernel module and begin to stop itself and the other server children. In some cases (i.e. for a position data request) the child is not to send the message to the kernel module, and then it answers independently to the client. Depending on the first field, the message can contain a data request, a simple order, or a more complex mission request. The message is completely decoded from the

RT-FIFO handler thread, that attempts to interpret the message and provides the data contained to the mission manager.

7   Timing Accuracy Experiments

When using a Real Time Operating System, it is very important to measure the accuracy of timing. We decided to make these measures via software building an assembler macro to read clock cycles every time an interruption is called. Hence, it was possible to obtain high-accuracy measures for sampling time and thread length, without introducing disturbances in the system.

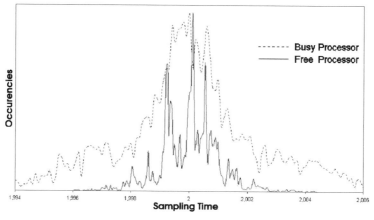

Fig. 15. Sampling time distribution with free and busy processor (properly scaled).

Figure 15 reports sampling time distributions obtained in two different conditions; in the first one, only the control threads are running on the computer, in the second one there are two programs performing complex mathematical tasks. As we can see, accuracy on sampling time is very high in both situations; as expected, covariance in the second one in higher than in the first one. Note that the introduction of a relative offset in the starting times of the different tasks is of paramount importance to obtain good results.

8   Conclusion

Having to tackle real problems is always a great source of experience. From a scientific point of view, the realization of the present version of SuperMARIO gave us the opportunity to design and develop a real-time architecture, to analyse its behavior and provide the lab with a really open set-up suitable for testing any kind of control algorithm. The architecture includes perhaps all the primitives needed to build a control system. In particular, the real-time routines can use different sampling time to optimize the CPU allocation, communication protocols have been defined between the real-time parts and the "user" (non real time) parts as well as between the latter and another computer. However, the greatest success was probably from an educational standpoint. Indeed the students involved in this work had the opportunity to follow a whole project from scratch. They learned a lot about designing cards, interfacing, tailoring a small footprint operating system, shaping loops taking friction into account and, most importantly, assembling all the parts together in one working system.

On the other hand, the project required a lot of time, as can be easily understood. The mechanical part had to be redesigned to get improved stiffness. The software too underwent dramatic modifications with time: the first release was indeed written under DOS. Also the interface card required a lot of study and subsequent attempts. Viceversa, the cost of the prototype was low even when compared with the basic commercial units. Obviously, our unit lacks range finders and high-level software; the former ones can be added with a small expense, the latter is not required at the moment and will be, in case, another opportunity of study.

References

1. B. Armstrong-Hélouvry, "Stick slip and control in low-speed motion," *IEEE Trans. on Automatic Control*, vol. 38, pp. 1483–1496, 1993.
2. A. Barabanov, *A Linux-Based Real-Time Operating System*, Master Thesis, New Mexico Institute of Mining and Technologies, Socorro, New Mexico, 1997.
3. E. Galvan, A. Torralba, and L.G. Franquelo, "A simple digital tachometer with high precision in a wide speed range," *Proc. of 20th IEEE Int. Conf. on Industrial Electronics, Control and Instrumentation*, pp. 920–923, 1994.
4. R.W. Hamming, *Digital Filters*, Prentice-Hall, 1989.
5. A. Macchelli, C. Melchiorri, and D. Pescoller "An experimental set-up for robotics and control system research using Real-Time Linux and Comau SMART 3-S robot," *Real-Time Linux Workshop*.
6. A.V. Oppenheim and R.W. Schafer, *Digital Signal Processing*, Prentice-Hall, 1975.
7. G. Oriolo, A. De Luca, and M. Vendittelli, "WMR control via dynamic feedback linearization: Design, implementation and experimental validation," *IEEE Trans. on Control Systems Technology*, vol. 10, pp. 835–852, 2002.
8. G. Oriolo, S. Panzieri, and G. Ulivi, "An iterative learning controller for nonholonomic mobile robots," *Int. J. of Robotics Research*, vol. 17, pp. 954–970, 1998.
9. J.I. Ripoll, *Tutorial de RTLinux*, http://bernia.disca.upv.es/rtportal.
10. W.R. Stevens, *UNIX Network Programming Vol. 1*, Prentice-Hall, 1998.
11. *Getting Started with RTLinux*, FSMLabs Inc., 2001.
12. *Real-time Programming in RTLinux*, FSMLabs Inc., 2002.
13. Web site contaning the RTLinux "Manifesto": http://www.rtlinux.org.
14. Web site of RTLinux authors: http://www.fsmlabs.com.

Coordination and Control of Multiarm Nonholonomic Mobile Manipulators

Giuseppe Casalino and Alessio Turetta

Dipartimento di Informatica Sistemistica e Telematica
Università di Genova
Via Opera Pia 13, 16145 Genova, Italy
<casalino,turetta>@dist.unige.it
http://www.graal.dist.unige.it

Abstract. This chapter deals with the problem of suitably coordinating the manoeuvring of a nonholonomic vehicle and the motion of a supported manipulation system (composed by one or two arms) when the overall system is commanded to execute a given grasping or manipulation task. The goal is that of suitably exploiting the extra degrees of freedom offered by the vehicle for better accomplishing the assigned task in a cooperative way.

1   Introduction

In the robotic literature, the field of mobile manipulators (i.e. a standard manipulator mounted on a mobile base or vehicle) has received a certain amount of attention since the beginning of the nineties, with the obvious objective of suitably exploiting the extra degrees of freedom offered by the vehicle for accomplishing specific (typically "long range") manipulation tasks that otherwise could not be executed completely.

Preliminary works in the field first focused on off-line motion planning of the overall structure [7,22,23], others focused on dynamic control with respect to pre-planned overall motions [11,16,12,25,21], while some others focused on kinematic and dynamic analysis only [26,24].

Moreover, within many of the early works, the manipulation and locomotion co-ordination problem was approached by assuming sequential motions of the platform and the manipulator (i.e. an approach phase performed via base motion only, then manipulation performed by the arm only).

On the other hand, to the best of authors' knowledge, one of the first papers where the reactive simultaneous coordination of locomotion and manipulation was proposed and preliminary developed dates to [27]. In this work a planar locomotion platform (unicycle-like) and a 2-dof manipulation structure are considered, while the concepts of manipulability ellipsoid and manipulability measure, see [30,31,29] and [19]) was explicitly used for assigning to the manipulator a so called "preferred posture", corresponding to the maximum level of its manipulability measure (MM). Then the manipulator was independently joint controlled, just in order to maintain such posture. In this condition, an additional joint velocity command (translating a desired absolute linear velocity for the end effector of the arm seen as a fixed-base one) was superimposed at joint level, thus inducing the arm to go slightly out from

B. Siciliano et al. (Eds.): Advances in Control of Articulated and Mobile Robots, STAR 10, pp. 171–190, 2004.
© Springer-Verlag Berlin Heidelberg 2004

its controlled posture, and in turn resulting in a drop of its actual *MM*. Such *MM* drop (or equivalently its corresponding posture mismatch) was finally compensated via the addition, to the arm base located on the platform, of a suitably evaluated linear velocity provided by the supporting vehicle itself.

Such scheme could work for both holonomic and nonholonomic vehicles (in the latter case only provided that the arm base was not located at the vehicle rotation pivot [27]) and clearly resulted in an overall structure where its composing entities (vehicle and manipulator), though separately controlled, acted in a simultaneous cooperative fashion.

A similar approach, extended to the case of multi-manipulator 3D systems, even if mounted on planar holonomic vehicles, was later successfully proposed in [20]. Following [27,20], many other works (see e.g. [2,4,3,28,5,6]) approached the locomotion and manipulation (simultaneous) coordination problem by explicitly keeping into account the Jacobian matrix of the overall structure, i.e. vehicle plus manipulator seen as a unique enlarged robotic structure, and extending the concept of manipulability ellipsoid and related *MM*. Then, the task-priority based control technique (originally introduced in [19] for fixed-base manipulators) could in turn be easily applied, in particular, by considering the singularity avoidance of the overall structure as a secondary task [5] with respect to the primary task of tracking the desired absolute end-effector motion.

Concerning successive approaches proposed in the literature —notwithstanding their theoretical framework allowing mobile manipulators to be substantially treated as analogous to fixed-base ones— it is the authors' opinion that they suffer from some drawbacks of both theoretical and practical nature. More specifically: a) such global approaches cannot be easily extended to the case of multiple manipulators supported by the same moving platform; b) they cannot easily respond to the increasing demand for modularity (functional, algorithmic, and Hw/Sw) within scalable complex robotic systems.

Motivated by the above considerations, but still inspired by the formerly mentioned work [27,20], the present work aims at proceeding further on the development of a general coordination theory for independently controlled vehicle and manipulators that naturally extends till the more complex cases of supported 3D multi-manipulators and 3D nonholonomic vehicles too, while always preserving modularity and scalability within the overall system.

The present chapter is organized as follows: in Section 2, the basic sub-problem of controlling the end effector of a fixed-base single arm while also avoiding its singularities is carefully reviewed, since it represents the fundamental basis for all successive developments. In Section 3 the problem of coordinating locomotion and manipulation for a single 3D arm and a 3D nonholonomic vehicle is developed within a fairly more general framework than that considered within [27]. Then, in Section 4, the previously obtained results are extended to the more general case of still 3D and nonholonomic multiarm mobile systems, when performing grasping and object manipulation tasks. Some conclusions and directions for future research activities are given in a final section.

2 Control of a Fixed-Base Single Arm with Singularity Avoidance

Let us consider a redundant fixed-base single arm, i.e. with a number of degrees of freedom (dof's) greater than six. Without loss of generality, let us refer to the arm in Fig. 1, where the wrist be constituted by a 3-dof rotational joint, typically of Euler and/or Roll-Pitch-Yaw type.

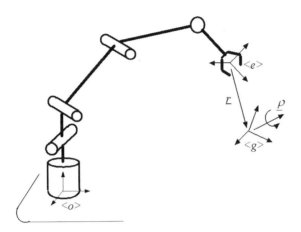

Fig. 1. A common example of a redundant fixed-base single arm.

In the figure, frame $< g >$ represents the "goal frame", which has to be reached (in position and orientation) by the "end-effector frame" $< e >$ of the manipulator. Let

$$e := [\, \rho^T \quad d^T \,]^T \tag{1}$$

be the collection of the misalignment error vector ρ and distance error vector d of frame $< e >$ with respect to <g>, when projected on world frame $< 0 >$. Also, let

$$\dot{x} := [\, \Omega^T \quad \nu^T \,]^T \tag{2}$$

be the collection of angular and linear velocities of $< e >$, still projected on $< 0 >$, where a small abuse of notation has occurred as for the use of the derivative. Consider the candidate Lyapunov function and its derivative

$$V := \frac{1}{2} e^T e \quad \Rightarrow \quad \dot{V} = -e^T \dot{x}. \tag{3}$$

It is easy to see that a choice of joint velocities satisfying at all times the condition

$$\dot{x} = \dot{\bar{x}} := \gamma e = J\dot{q} \qquad \gamma > 0, \tag{4}$$

where the upper bar denotes the reference value and $J(q)$ is the Jacobian matrix of $<e>$ with respect to $<0>$, would drive $<e>$ toward $<g>$ asymptotically. As it is well known, Eq. (4) must be solved in real time for the joint velocities via regularized Jacobian matrix pseudo-inversion (see e.g. the damped least-squares inverse in [19]), in order to prevent joint velocities to grow toward infinity in the vicinity of any Jacobian singularity that could be encountered during the arm motion. The net effect of the regularization is that of progressively reducing to zero those components of the joint velocity vector that otherwise would unacceptably grow to infinity.

Despite such benefit, some drawbacks are implied by the regularization itself, e.g. unpredictable motion perturbations generally occurring whenever crossing singularities, or even possibilities of getting stuck in correspondence of certain configurations with the need of complex manoeuvring for departing.

Therefore, in order to reduce the chances for such occurrences, a secondary task, attempting to maintain the arm far from singularities while accomplishing the primary task (4), should actually be introduced and executed by exploiting redundancy of the arm. To this end, by referring to the so-called "manipulability measure" (*MM*) [19,29], i.e. the scalar quantity

$$\mu := \det(JJ^T) \geq 0 \qquad (5)$$

and then considering its time derivative

$$\dot{\mu} = p^T \dot{q} \quad ; \quad p := \frac{\partial \mu}{\partial q} \qquad (6)$$

where the row vector p^T (always non-zero in correspondence of any $\mu > 0$) can be efficiently evaluated in real time via the procedure developed in [18], we recognize that a choice of the joint velocities aimed at satisfying also the condition

$$\dot{\mu} = \dot{\bar{\mu}} := \lambda\mu \qquad \lambda > 0 \qquad (7)$$

would possibly and sensibly reduce the risk of singularity occurrence during motion, provided that the starting position be far from singularities.

Conditions (4) and (7) respectively represent the so-called primary task (i.e. end-effector reaching $<g>$) and secondary task (i.e. arm attempting to remain far from singularities) which directly lead to the following expression for the joint velocities [19]:

$$\dot{q} = J^{\#}\dot{\bar{x}} + h(\dot{\bar{\mu}} - k\dot{\bar{x}}) \qquad (8)$$

with

$$k := pJ^{\#} \qquad (9)$$

and

$$h := \left[p\left(I - J^{\#}J\right)\right]^{\#} \qquad (10)$$

where all matrix pseudo-inversions are assumed to be performed in regularized form.

The vector h in (10) is proved to belong to the null space of J ($N(J)$) in correspondence of any arm posture such that $\mu \geq \mu^*$, being μ^* the a priori assigned *MM* threshold below which the regularization embedded in the pseudo-inversion of Jacobian matrix J is made active (see Appendix A).

Though the adoption of (8) reduces the chances for singularity occurrences, while accomplishing the desired end-effector motion, such risk cannot be completely avoided via the use of the sole solution (8). In fact (see Appendix B), such risk might occur even for cases where the end-effector motions are required to completely lie within the so called dexterous reachable workspace (*DRW*) (i.e. the simply connected subset of the arm workspace where any end-effector attitude can be assigned via arm postures admitting a non-empty subset such that $\mu > \mu^*$). On the other hand, whenever the goal frame $< g >$ is located outside the *DRW*, such risk obviously becomes unavoidable (see Appendix B).

As a consequence, the need for exchanging (possibly in a smooth way) the priority order between the two tasks naturally arises whenever an incoming risk of singularity occurrence is foreseen. To this extent, a nice approach has been recently proposed in [18], as an important extension of the works [8,9] on the subject. The idea is quite simple: first of all a minimum value $\mu_0 > \mu^*$ for *MM* is a priori established, beyond which the actual μ is desired to stay, which in turn induces a restriction on the originally defined *DRW*); then, during motion, μ is continuously monitored and, if lower than μ_0 in the form $\mu^* \leq \mu < \mu_0$ (then possibly also at the starting configuration) the Cartesian velocity reference $\dot{\bar{x}}$ is corrected as

$$\dot{x}^* = \dot{\bar{x}} + \dot{z} \tag{11}$$

where the additional signal $\dot{z}$ has to be chosen (if possible) in such a way that

$$\dot{\bar{\mu}} - k\dot{\bar{x}} - k\dot{z} = 0 \tag{12}$$

must hold.

It is easy to see that replacing $\dot{\bar{x}}$ with $\dot{x}^*$ in (11) and modifying $\dot{q}$ in (8) accordingly, condition (7) turns out to be exactly satisfied, which implies a progressive increase of μ toward μ_0 (regardless of being $< g >$ located inside or outside *DRW*), while also meaning an implicit exchange of priority order between the two tasks. Also observe that, being (12) a scalar condition, it consequently admits ∞^5 solutions in the correction vector $\dot{z}$, among which the following (minimum norm) is certainly the most suitable one for the case of a fixed-base single arm considered in this section, i.e.

$$\dot{z} = k^{\#}(\dot{\bar{\mu}} - k\dot{\bar{x}}) \tag{13}$$

At this point, while referring to Appendix C for some additional comments concerning condition (12) and related solution (13), we can conclude this section by simply noting that, in order to be comprehensive of the overall cases $\mu^* < \mu < \mu_0$ or $\mu^* < \mu > \mu_0$, while also avoiding any possible chattering in the vicinity of the threshold value μ_0, it is actually always convenient to adopt the following expression for $\dot{z}$:

$$\dot{z} = (\alpha k)^{\#}(\dot{\bar{\mu}} - k\dot{\bar{x}}) \tag{14}$$

being $\alpha(\mu)$ a continuous scalar function of μ, which is unitary for $\mu \leq \mu_0$ and bell-shaped, tending to zero within a finite support for $\mu > \mu_0$. Obviously enough, with such final adjustment, the smooth transition between the two different cases of task priority turns out to be automatically guaranteed.

3 Control of a Single-Arm Nonholonomic Mobile Manipulator

The case of a redundant arm mounted on a 3D moving base as in Fig. 2 is now considered. The vehicle is assumed to nonholonomic, in the sense that it allows a linear velocity vector $\underline{\nu}$ only directed along the principal vehicle axis, and an angular velocity vector $\underline{\Omega}$ only lying on a plane passing through a known point of such principal axis, and orthogonal to it. The arm and the vehicle are regarded

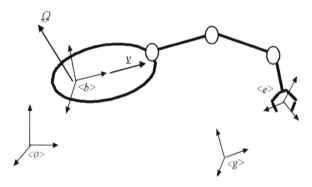

Fig. 2. Single arm supported by a vehicle.

as two separate "basic robotic units", whose motions however needs to be suitably coordinated for the execution of a common task (i.e. making again $< e >$ converge toward $< g >$) to be realized in a cooperative way. In this context, the arm is assumed to be separately controlled by a control law structurally identical to the one of the previous section, with the noticeable addition of an external signal ζ to be used by an appropriate upper layer for coordination purposes. More specifically, this is achieved by imposing the Cartesian velocity reference in (11) to attain the more general form

$$\dot{x}^* = \dot{\hat{x}} + \dot{z} \tag{15}$$

with

$$\dot{\hat{x}} := \dot{\bar{x}} + \dot{\zeta} \tag{16}$$

which in turn implies that $\dot{z}$ is to match the new reference signal $\dot{\hat{x}}$; hence, $\dot{z}$ is chosen so as to satisfy

$$\ddot{\bar{\mu}} - k\dot{\hat{x}} - k\dot{z} = 0 \tag{17}$$

when $\mu \leq \mu_0$, or is set to zero otherwise.

With the above considerations in mind, let us now approach the overall control problem of having $<e>$ converging to $<g>$ by again considering the candidate Lyapunov function (3). Its time derivative now takes on the form

$$\dot{V} = -e^T(\dot{x} + \dot{X}) \tag{18}$$

being $\dot{x}$ and $\dot{X}$ the contributions to the end-effector motion separately produced by the arm and the vehicle, respectively, both projected on world frame $<0>$. More specifically, for $\dot{X}$ we actually have

$$\dot{X} = S\dot{\theta} \tag{19}$$

where $\dot{\theta}$ is the three-dimensional vector resulting from the collection of the two non-null components w of Ω and the sole non-null component u of ν, provided that both Ω and ν are projected on the vehicle fixed frame $<b>$ as indicated in Fig. 2; that is

$$\dot{\theta} = [\, w^T \quad u\,]^T. \tag{20}$$

Also in (19) the matrix

$$S = HQ \tag{21}$$

where is a (6×6) matrix representing the instantaneous rigid-body velocity transformation from vehicle frame $<b>$ to the end-effector frame $<e>$ (input velocities projected on $<b>$, output velocities projected on world frame $<0>$), while Q is simply a full-rank (6×3) selection matrix, suitably composed by 0 and 1 elements. Notice that the (6×3) matrix S is also full-rank.

Folding (19) into (18) gives

$$\dot{V} = -e^T(\dot{x} + S\dot{\theta}). \tag{22}$$

At this point, by choosing the Cartesian reference velocity $\dot{x}$ in (22) as in the form (15) and (16) yields

$$\dot{V} = -e^T(\dot{\bar{x}} + \dot{z} + S\dot{\theta}) = -e^T\left[(\dot{\bar{x}} + \dot{\zeta}) + \dot{z} + S\dot{\theta}\right]. \tag{23}$$

Now, let us express the coordination signal $\dot{\zeta}$ in a form just opposite to (19); i.e.

$$\dot{\zeta} = -S\dot{\theta} \tag{24}$$

with $\dot{\theta}$ to be assigned also to the supporting vehicle. Further, let us choose $\dot{\zeta}$, if possible, in such a way to satisfy a condition similar to (12), i.e.

$$\dot{\bar{\mu}} - k\dot{\bar{x}} + kS\dot{\theta} = 0 \tag{25}$$

when $\mu \leq \mu_0$, to be zero otherwise. Then, under the above assumptions it is not difficult to realize that the following two facts must necessarily hold.

a) The internal signal $\dot{z}$ turns out to be identically zero since its role is completely accomplished by the coordinating signal $\dot{\zeta}$ itself.
   In fact, when $\mu > \mu_0$ both $\dot{z}$ and $\dot{\theta}$ (and then also $\dot{\zeta}$) are chosen to be zero; while for $\mu \leq \mu_0$, from (25) we have, also keeping into account (24) and (16),

$$\dot{\bar{\mu}} - k\dot{\hat{x}} = 0 \tag{26}$$

thus implying the internal condition (17) to be naturally satisfied by $\dot{z} = 0$

b) Due to the specific structure (24) assigned to $\dot{\zeta}$, its contribution to the end-effector motion is compensated by the opposite motion contribution provided by the vehicle.

Then, as a consequence of the above two facts, it follows that expression (23) actually takes on the form

$$\dot{V} = -e^T \dot{\bar{x}} < 0 \tag{27}$$

which in fact guarantees the convergence of the end-effector frame $< e >$ toward the goal frame $< g >$ without any restriction.

At this point, in order to satisfy condition (25) when $\mu \leq \mu_0$, we note that it certainly admits ∞^2 solutions for $\dot{\theta}$, provided we do not fall within the very unlikely singularity characterized by having vector k^T orthogonal to the range space of S ($R(S)$); for the time being a detailed analysis of such event is however out of the scopes of this chapter. It follows that a suitable choice for $\dot{\theta}$ (i.e. the minimum norm one, requiring a minimal vehicle motion when $\mu \leq \mu_0$) is

$$\dot{\bar{\theta}} = -(kS)^{\#}(\dot{\bar{\mu}} - k\dot{\bar{x}}) \tag{28}$$

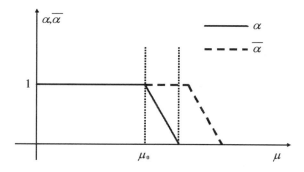

Fig. 3. Relationship between α and $\overline{\alpha}$.

It should be emphasized how the proposed coordination law does not actually require —apart the addition of the external coordination command $\dot{\zeta}$— any intervention on the functional and algorithmic structure of the manipulator control system, which remains the same as for the fixed-base case. Moreover note that, since the internal functionality concerning the generation of $\dot{z}$ remains always active (though producing a null signal), it naturally reduces to a sort of "safety functionality": ready to automatically come into play whenever any sort of coordination failure occurs. Finally, in order to avoid any possible chattering around the manipulability threshold μ_0, we require that the signal $\dot{\zeta}$ should actually be generated via the smoothed form

$$\dot{\overline{\theta}} = -(\overline{\alpha}kS)^{\#}(\dot{\overline{\mu}} - k\dot{\overline{x}}) \tag{29}$$

where $\overline{\alpha}(\mu)$ is again a continuous scalar function of μ, which is unitary for $\mu \leq \mu_0$ and bell-shaped, tending to zero within a finite support for $\mu > \mu_0$. Moreover note that, in order to also avoid any possible interference between $\dot{\zeta}$ —now smoothly generated via (29), (24) and $\dot{z}$ which is maintained active via the smoothed form (14))— we should furtherly shape $\overline{\alpha}(\mu)$, with respect to $\alpha(\mu)$, in such a way as to be certainly unitary within the whole finite support where $\alpha(\mu) > 0$ (see Fig. 3).

As it can be easily realized, with such a choice for $\overline{\alpha}$ and α, signal $\dot{z}$ is always null, even during the smooth transition phase of $\dot{\overline{\theta}}$ (and then $\dot{\zeta}$).

4   Control of a Dual-Arm Nonholonomic Mobile Manipulator

The results of the previous section are hereafter extended to the case of a dual-arm nonholonomic mobile manipulator of the type of Fig. 4, when performing grasping operations. As a matter of fact, a grasping operation to be performed by the overall system simply corresponds to the global task of having the two end-effector frames $< e_1 >$, $< e_2 >$ asymptotically converging to the goal frames $< g_1 >$, $< g_2 >$ respectively (Fig. 4), while obviously maintaining the desired minimum level of manipulability for each arm.

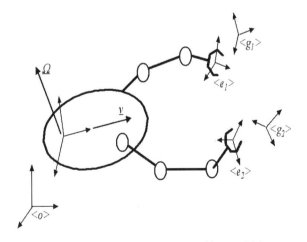

Fig. 4. Dual arm system supported by a vehicle.

By still assuming each arm to be separately controlled, as done in the previous section, let us start again by considering the following global candidate Lyapunov function, with an obvious meaning of the introduced terms,

$$V := \frac{1}{2} \left(e_1^T e_1 + e_2^T e_2 \right) \tag{30}$$

whose time derivative is

$$\dot{V} = -e_1^T (\dot{x}_1 + \dot{X}_1) - e_2^T (\dot{x}_2 + \dot{X}_2) := -e^T (\dot{x} + \dot{X}). \tag{31}$$

Then, by performing the same analysis leading to (23) in the previous section, we get

$$\dot{V} = -e^T (\dot{x}^* + S\dot{\theta}) = -e^T (\dot{\hat{x}} + \dot{z} + S\dot{\theta}) = -e^T \left[(\dot{\bar{x}} + \dot{\zeta}) + \dot{z} + S\dot{\theta} \right] \tag{32}$$

where now

$$\begin{aligned}
\dot{x}^* &:= [\, \dot{x}_1^{*T} \quad \dot{x}_2^{*T} \,]^T \\
S &:= [\, S_1^T \quad S_2^T \,]^T \\
\dot{\hat{x}} &:= [\, \dot{\hat{x}}_1^T \quad \dot{\hat{x}}_2^T \,]^T \\
\dot{z} &:= [\, \dot{z}_1^T \quad \dot{z}_2^T \,]^T \\
\dot{\bar{x}} &:= [\, \dot{\bar{x}}_1^T \quad \dot{\bar{x}}_2^T \,]^T \\
\dot{\zeta} &:= [\, \dot{\zeta}_1^T \quad \dot{\zeta}_2^T \,]^T
\end{aligned} \tag{33}$$

The (12×6) matrix S is still of full column rank type, and the overall coordination signal $\dot{\zeta}$ is to be suitably chosen. To this end, on the basis of considerations analogous to those in the previous section, provided we can still preserve *MM* for both arms via the external coordination signal

$$\dot{\zeta} = -S\dot{\theta} \tag{34}$$

with $\dot{\theta}$ to be assigned to the vehicle too, we have then

$$\dot{z} = 0 \tag{35}$$
$$\dot{V} = -e^T \dot{\bar{x}} < 0 \tag{36}$$

guaranteeing the accomplishment of the assigned grasping task.

Then, in order to verify whether *MM* can be still maintained within the desired levels via $\dot{\zeta}$, let us analyze the corresponding four possible cases:

a) $\mu_1 > \mu_0$; $\mu_2 > \mu_0$

In this case, since *MM* is adequate for both arms, we must obviously set

$$\dot{\zeta} = 0 \Rightarrow \dot{\theta} = 0 \tag{37}$$

b) $\mu_1 > \mu_0$; $\mu_2 \leq \mu_0$

In this case, since *MM* must be recovered only for arm 1, this requires the fulfillment of its relevant condition (25), i.e. by still looking for a minimum norm solution in $\dot{\theta}$)

$$\dot{\bar{\mu}}_1 - k_1 \dot{\bar{x}}_1 + k_1 S_1 \dot{\theta} = 0 \quad \Leftrightarrow \quad \dot{\theta} = -(k_1 S_1)^{\#}(\dot{\bar{\mu}}_1 - k_1 \dot{\bar{x}}_1) \tag{38}$$

which unavoidably induces (though not necessary) a correction term also on arm 2; i.e. the term

$$\dot{\zeta}_2 = -S_2 \dot{\theta} \tag{39}$$

which can be anyhow accepted by arm 2 itself, since its *MM* is greater than the minimum threshold μ_0.

c) $\mu_1 \leq \mu_0$; $\mu_2 > \mu_0$

This is simply the dual of the previous case, thus leading to

$$\dot{\bar{\mu}}_2 - k_2 \dot{\bar{x}}_2 + k_2 S_2 \dot{\theta} = 0 \quad \Leftrightarrow \quad \dot{\theta} = -(k_2 S_2)^{\#}(\dot{\bar{\mu}}_2 - k_2 \dot{\bar{x}}_2) \tag{40}$$

analogously implying, unavoidably but acceptably

$$\dot{\zeta}_1 = -S_1 \dot{\theta} \tag{41}$$

d) $\mu_1 \leq \mu_0$; $\mu_2 \leq \mu_0$

In this case, since *MM* must be recovered for both arms, this requires the contemporary fulfillment of condition (25), i.e.

$$\begin{cases} \dot{\bar{\mu}}_1 - k_1 \dot{\bar{x}}_1 + k_1 S_1 \dot{\theta} = 0 \\ \dot{\bar{\mu}}_2 - k_2 \dot{\bar{x}}_2 + k_2 S_2 \dot{\theta} = 0 \end{cases} \tag{42}$$

or in a more compact notation

$$\dot{\bar{\mu}} - K\dot{\bar{x}} + KS\dot{\theta} = 0 \tag{43}$$

where obviously

$$\dot{\bar{\mu}} := \begin{bmatrix} \dot{\bar{\mu}}_1^T & \dot{\bar{\mu}}_2^T \end{bmatrix}^T \tag{44}$$
$$K := \mathrm{diag}\,(k_1, k_2)\,.$$

Then, by noting that (43) actually admits ∞^1 solutions, provided that remains full row rank, we can choose the minimum norm solution for $\dot{\theta}$, that is

$$\dot{\theta} = -(KS)^{\#}(\dot{\bar{\mu}} - K\dot{\bar{x}})\,. \tag{45}$$

Notice that, as it concerns the full rankness of matrix KS —notwithstanding the fact a thorough analysis is outside of the scopes of the present work— we can devise, according to intuition, at least one case where full rankness of KS is certainly lost; this simply corresponds to the case where goal frames $< g_1 >$, $< g_2 >$ are located at the opposite edges of the vehicle, and quite far from it.

Also notice that whenever the overall system is, for some reasons, made to tend to such (unreasonable) configurations, then $\dot{\overline{\theta}}$ in (45) naturally tends to zero (and consequently ζ too) due to the assumed embedded regularization within the pseudoinversion of KS. This will consequently make the internal "safety" correction term $\dot{z}$ come into play for still separately guaranteeing manipulability of each arm, while the vehicle will gradually stops its motion. Obviously enough, the assigned grasping task (being an impossible one) will therefore not be at all accomplished.

Finally notice that for the same reasons mentioned in the previous section, also in this case $\dot{\theta}$ should be generated via the smooth form

$$\dot{\overline{\theta}} = -(\overline{\alpha}KS)^{\#}(\dot{\overline{\mu}} - K\dot{\overline{x}}) \tag{46}$$

where now

$$\overline{\alpha} = \text{diag}\,(\overline{\alpha}_1, \overline{\alpha}_2) \tag{47}$$

with $\overline{\alpha}_1$, $\overline{\alpha}_2$ having the same shape as $\overline{\alpha}$ in Fig. 3.

5   Object Manipulation via Dual-Arm Nonholonomic Mobile Manipulator

The results obtained in the previous section will be now easily extended to the case of an object manipulated by a dual-arm mobile nonholonomic system, as depicted in Fig. 5.

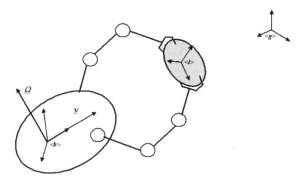

Fig. 5. Object manipulated by the system.

The manipulated (lightweight) object is assumed to be firmly grasped by the end-effector of the dual-arm system. The object itself is characterized by its own fixed body frame $< l >$ which is required to be asymptotically convergent toward an assigned goal frame $< g >$.

By denoting with e_l the generalized error (position and orientation) of frame $< l >$ with respect to $< g >$, let us define as

$$\dot{\overline{x}}_l := \gamma e_l \quad ; \quad \gamma > 0 \tag{48}$$

the velocity reference signal that, once applied to $< l >$, would guarantee $< l >$ itself to be asymptotically convergent to $< g >$. With this in mind, let us also assume the commanded Cartesian velocity vector for the two end effectors to be now of the form

$$\dot{x}^* = \dot{\hat{x}} + \dot{z} = \left(P\dot{\overline{x}}_l + \dot{\zeta} \right) + \dot{z} = \left(P\dot{\overline{x}}_l - S\dot{\theta} \right) + \dot{z} \tag{49}$$

where

$$P := [P_1^T \quad P_2^T]^T \tag{50}$$

is the collection of the velocity rigid-body transformation matrices P_1 e P_2 from frame $< l >$ to $< e_1 >$ and $< e_2 >$, respectively, while the other terms remain the same as in the previous section. As a consequence, we have that signal $\dot{\theta}$, though being now slightly modified (compare with (46)) as

$$\dot{\theta} = -(\overline{\alpha}KS)^{\#}(\overline{\mu} - KP\dot{\overline{x}}_l) \tag{51}$$

will again force the internal signal $\dot{z}$ to satisfy the zeroing condition

$$\dot{z} = 0 \tag{52}$$

At this point, by explicitly keeping (52) into account, we can consequently note the full compatibility of the resulting $\dot{x}^*$ with the assumed grasping constraints, that is the fulfilment of conditions

$$P_1^{-1}\left(P_1\dot{\overline{x}}_l \right) = P_2^{-1}\left(P_2\dot{\overline{x}}_l \right) = \dot{\overline{x}}_l \tag{53}$$

and

$$P_1^{-1}\left(-S_1\dot{\theta} \right) = P_2^{-1}\left(-S_2\dot{\theta} \right) := -\overline{S}\dot{\theta} \quad \forall \dot{\theta} \tag{54}$$

being $\overline{S}$ the resulting overall rigid-body velocity transformation matrix from vehicle frame $< b >$ to object frame $< l >$.

Then we can conclude that the dual-arm velocity contribution $\dot{x}_l$ takes on the form

$$\dot{x}_l = \dot{\overline{x}}_l - \overline{S}\dot{\theta} \tag{55}$$

which, upon addition of the velocity contribution $\dot{X} = \overline{S}\dot{\theta}$ provided by the vehicle, in turns leads to the following expression for the object overall absolute velocity

$$\left(\dot{x}_l + \dot{X}_l \right) = \dot{\overline{x}}_l. \tag{56}$$

This obviously guarantees the desired asymptotic convergence of $< l >$ toward $< g >$.

6 Simulation Results

In order to validate the proposed coordination method, some preliminary simulations have been carried out, though they are referred to the intermediate case of a single arm mounted on a nonholonomic vehicle. More specifically, a mobile manipulator composed by a 7-dof arm mounted on a 3D nonholonomic base is considered. The end effector of the arm is asked to reach a Cartesian position located sufficiently far form the starting one, without changing its original orientation.

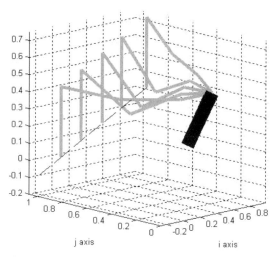

Fig. 6. First part of the system motion: only arm moving.

Figures 6, 7 refer to the initial part of the system motion, characterized by having $\mu = 0.6 > \mu_0 = 0.4$ at the initial time; note that the black box in Fig. 6 represents the 3D mobile base. During this motion part, *MM* first increases under the action of the secondary task, but then it starts to decrease, due to the persistency of the primary task. During this period, the vehicle remains fixed in its original position, while the $\bar{\alpha}$ parameter obviously maintains its original null value.

Nevertheless, once *MM* crosses from above the established activation threshold 0.6 for the $\bar{\alpha}$ parameter (remember Fig. 3 and see Fig. 7), $\bar{\alpha}$ itself increases toward unity (while μ reduces toward 0.4), thus causing the vehicle to move (in order to compensate for the Cartesian extra command signal ζ now added at the arm level) whereas the end effector continues its unperturbed motion toward the requested final position. Also notice how *MM* always remains above the minimal threshold $\mu_0 = 0.4$ (still see Fig. 3) while starting again to increase when the task is almost completed, thus causing $\bar{\alpha}$ reducing again to zero while the vehicle ends its motion too.

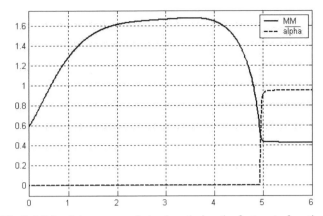

Fig. 7. *MM* and $\dot{\alpha}$ parameter behaviors during the first part of motion.

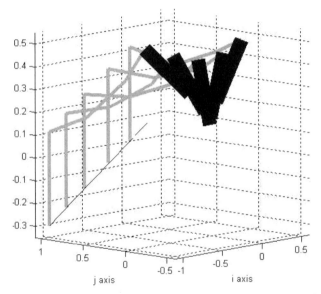

Fig. 8. Second part of the system motion: both arm and vehicle moving.

7   Conclusion

This chapter has considered the problem of devising control strategies for the continuous (smooth) coordination of the motions of nonholonomic vehicles and manipulation structures, whenever the latter are mounted on the vehicle to the aim of exploiting the overall resulting redundant structure for executing manipulation tasks. Since the mounted arms, as well as the vehicle, have been regarded as a set of independently controlled "basic robotic units", the coordination problem has been reduced to the real-time generation of the coordination signals for the underlying structures, which in turn allows the coordinated smooth accomplishment of the over-

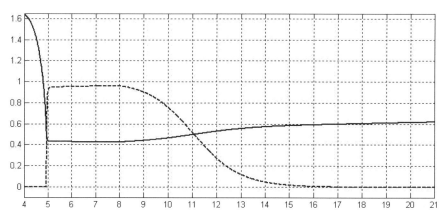

Fig. 9. *MM* and $\dot{\alpha}$ parameter behaviors during the second part of motion.

all assigned task. To this end a fundamental role has been played by the concept of Manipulability Measure and the efficient techniques for controlling it.

Before concluding, it is worth noticing that the proposed control and coordination schemes fall within the so-called "resolved-rate" robot control techniques, that is the category of robot control methods where a pure kinematic nature is implicitly assumed for the underlying robotic structures. Such assumption, from a practical point of view, simply translates into the assumption of a perfect (or at least quasi-perfect) velocity control performed at the joint levels of the underlying structures, via suitable (inner and local) joint velocity control loops; this is the case, for instance, of exact or approximate computer torque methods, possibly coupled with "high gain" linear control feedbacks.

As a matter of fact, since all the dynamic control aspects actually remain confined within each single loop associated with the corresponding robotic structure, the adoption of such standpoint leads to an easier construction of modular control architectures for scalable complex robotic systems.

As a counterpart to the above-mentioned "resolved rate" approach, the full dynamic approach proposed in [15] stands as a nice extension of the "operational space" formulation for dynamic control of complex robotic systems [13,14]. Within such approach an operational space dynamic model is first obtained by projecting the whole dynamics in the operational space, while leaving the remaining part of dynamics within the null space associated with the redundant mechanisms. As a result, such two dynamic models turn out forming the basis for the dynamic coordination strategies considered in [13–15]. Apart from apparent major implementation complexities involved in the operational space formulation, an investigation aiming at relating the two approaches (i.e. resolved rate vs. operational space) seems to be still lacking yet, constituting a topic of further research; preliminary attempts to identify such relationships can actually be found in [1,10].

Acknowledgement

This work has been co-funded by ASI, within a special project devoted to "functional and algorithmic control architectures for space robots".

References

1. M. Aicardi, G. Cannata, and G. Casalino, "Stability and robustness analysis of a two-layered hierarchical architecture for the control of robots in the operational space," *1995 IEEE Int. Conf. on Robotics and Automation*, pp. 2771–2778, 1995.
2. G. Antonelli and S. Chiaverini, "Task-priority redundant resolution for underwater vehicle-manipulator systems," *Proc. of 1998 IEEE Int. Conf. on Robotics and Automation*, pp. 768–773, 1998.
3. G. Antonelli and S. Chiaverini, "Fuzzy redundancy resolution and motion coordination for underwater vehicle-manipulator systems," *IEEE Trans. on Fuzzy Systems*, vol. 11, pp. 109–120, 2003.
4. G. Antonelli and S. Chiaverini, "A fuzzy approach to redundancy resolution for underwater vehicle-manipulator systems," *Control Engineering Practice*, vol. 11, pp. 445-452, 2003.
5. B. Bayle, J.Y. Forquet, and M. Renaud, "Manipulability analysis for mobile manipulators," *Proc. of 2001 IEEE Int. Conf. on Robotics and Automation*, pp. 1251–1256, 2001.
6. B. Bayle, J.Y. Forquet, and M. Renaud, "Using manipulability with non-holonomic mobile manipulators," *Int. Conf. on Field and Service Robotics*, pp. 343–348, 2001.
7. W.F. Carriker, P.K. Khosla, and B.H. Krogh, "Path planning for mobile manipulators for multiple task execution," *IEEE Trans. on Robotics and Automation*, vol. 7, pp 403–408, 1991.
8. G. Casalino, D. Angeletti, T. Bozzo, and G. Cannata, "Strategies for control and coordination within multiarm systems," in S. Nicosia, B. Siciliano, A. Bicchi, and P. Valigi (Eds.), *RAMSETE — Articulated and Mobile Robots for SErvices and TEchnology*, pp. 1–26, Springer Verlag, 2001.
9. G. Casalino, D. Angeletti, T. Bozzo, and G. Marani "Dexterous underwater object manipulation via multirobot cooperating systems," *Proc. of 2001 IEEE Int. Conf. on Robotics and Automation*, pp. 3220–3225, 2001.
10. G. Casalino, G. Cannata, G. Panin, and A. Caffaz, "On a two-level hierarchical structure for the dynamic control of multifingered manipulation," *Proc. of 2001 IEEE Int. Conf. on Robotics and Automation*, pp. 77–84, 2001.
11. S. Dubowsky and W.A.B. Tanner, "A study of the dynamics and control of mobile manipulators subjected to vehicle disturbances," *Proc. of 1987 IEEE Int. Conf. on Robotics and Automation*, pp. 111–117, 1987.
12. N.A.M. Hootsmans and S. Dubowsky, "Large motion control of mobile manipulators including vehicle suspensions characteristics," *Proc. of 1991 IEEE Int. Conf. on Robotics and Automation*, pp. 2336–2341, 1991.
13. O. Khatib, "A unified approach to motion and force control of robot manipulators: The operational space formulation," *IEEE J. of Robotics and Automation*, vol. 3, pp. 43–53, 1987.
14. O. Khatib, "Inertial properties in robotics manipulation: An object-level framework," *Int. J. of Robotics Research*, vol. 14, pp. 19–36, 1995.

15. O. Khatib, K. Yokoi, K. Chang, D. Ruspini, R. Holmberg, and A. Casal, "Coordination and decentralized cooperation of multiple mobile manipulators," *J. of Robotic Systems*, vol. 13, pp. 755–764, 1996.

16. K. Liu and F. Lewis, "Decentralized continuous robust controller for mobile robots," *Proc. of 1990 IEEE Int. Conf. on Robotics and Automation*, pp. 1822–1826, 1990.

17. A.A. Maciejewski and C.A. Klein, "Obstacle avoidance for kinematically redundant manipulators in dynamically varying environments," *Int. J. of Robotics Research*, vol. 4, no. 3, pp. 109–117, 1985.

18. G. Marani, J. Kim, and J. Yuh, "A real-time approach for singularity avoidance in resolved motion rate control of robotic manipulators," *Proc. of 2002 IEEE Int. Conf. on Robotics and Automation*, pp. 1973–1978, 2002.

19. Y. Nakamura, *Advanced Robotics: Redundancy and Optimization*, Addison Wesley, 1991.

20. U.M. Nassal, "Motion cooordination and reactive control of autonomous multi-manipulator systems," *J. of Robotic Systems*, vol. 13, pp. 737–754, 1996.

21. C. Perrier, L. Cellier, P. Dauchez, P. Fraisse, E. Degoulange, and F. Pierrot, "Position/force control of a manipulator mounted on a vehicle," *J. of Robotic Systems*, vol. 13, pp. 687–698, 1996.

22. F.G. Pin and J.C. Culioli, "Multi-criteria position and configuration optimization for redundant platform/manipulator systems," *Proc. of IEEE Work. on Intelligent Robots and Systems*, pp. 103–107, 1990.

23. F.G. Pin, K.A. Morgansen, F.A. Tulloc, C.J. Hacker, and K.B. Gower, "Motion planning for mobile manipulators with a non-holonomic constraint using the FSP (Full Space Parameterisation) method," *J. of Robotic Systems*, vol. 13, pp. 723–736, 1996.

24. H. Seraji, "An on-line approach to coordinated mobility and manipulation," *Proc. of 1993 IEEE Int. Conf. on Robotics and Automation*, vol. 1, pp. 28–35, 1993.

25. K.A. Tahboub, "Robust control of mobile manipulators," *J. of Robotic Systems*, vol. 13, pp. 699–708, 1996.

26. Y. Yamamoto and X. Yun, "Coordinating locomotion and manipulation of a mobile manipulator," *Proc. of 31st IEEE Conf. on Decision and Control*, pp. 2643–2648, 1993.

27. Y. Yamamoto and X. Yun, "Coordinating locomotion and manipulation of a mobile manipulator," *IEEE Trans. on Automatic Control*, vol. 39, pp. 1326–1332, 1994.

28. Y. Yamamoto and X. Yun, "Unified analysis on mobility and manipulability of mobile manipulators," *Proc. of 1999 IEEE Int. Conf. on Robotics and Automation*, pp. 1200–1206, 1999.

29. T. Yoshikawa, "Analysis and control of robot manipulators with redundancy," in M. Brady and R. Paul (Eds.), *Robotics Research: The First International Symposium*, pp. 735–747, MIT Press, 1984.

30. T. Yoshikawa, "Manipulability of robotic mechanisms," *Int. J. of Robotics Research*, vol. 4, no. 1, pp. 3–9, 1985.

31. T. Yoshikawa, *Foundations of Robotics: Analysis and Control*, MIT Press, 1990.

Appendix A

Consider vector h by rewriting it in its expanded form, that is

$$h = \left(I - J^{\#}J\right) p^T \frac{1}{(\phi + \epsilon)} \tag{57}$$

where for ease of notation we have let

$$\phi = p\left(I - J^{\#}J\right)\left(I - J^{\#}J\right)p^{T} \tag{58}$$

and $\epsilon\left(\phi\right)$ represents the regularization factor for the involved inversion, i.e. a bell-shaped continuous scalar function of ϕ, attaining its (small) maximum ϵ_0 in correspondence of $\phi = 0$ and tending to zero within an a priori finite support ϕ^{*}.

As it can be easily verified, under the assumption $\mu \geq \mu^{*}$ (i.e. J full rank) expression (57) consequently reduces to the projection of vector p^{T} on $N(J)$, simply normalized by the always non-zero coefficient $(\phi + \epsilon)$, with ϕ naturally coinciding with the squared norm of the projection of p^{T} itself on $N(J)$.

Appendix B

First, let us consider any desired end-effector motion trajectory totally evolving inside *DRW* (as it could be for instance the asymptotic goal reaching established by $\dot{\bar{x}} = \gamma e$ whenever both $< e >$ at the starting point and $< g >$ are located inside *DRW*). By definition of *DRW*, the end-effector position/orientation corresponding to any point along such motion trajectory could actually be obtained via a non-empty compact set of underlying arm postures having $\mu \geq \mu^{*}$. Then just assume $\mu \geq \mu^{*}$ in correspondence of the actual arm posture, and note from (8) that, after some simple algebra, we have

$$\dot{x} = \dot{\bar{x}} \tag{59}$$

$$\dot{\mu} = \left[\frac{\phi}{\phi + \epsilon}\right]\dot{\bar{\mu}} + \left[\frac{\epsilon}{\phi + \epsilon}\right]k\dot{\bar{x}}. \tag{60}$$

This clearly shows that, while the arm will nearby maintain its motion along the desired trajectory, the corresponding *MM* will unconditionally exhibit a non-decreasing behavior (i.e. the arm will also attempt to maintain μ above the minimum value μ^{*}) only if the current underlying posture (other than being such that $\mu \geq \mu^{*}$) will also continue to satisfy the condition $\phi \geq \phi^{*}$ (i.e. until a non-negligible projection of p^{T} on $N(J)$ exists, which implies $\epsilon = 0$, and consequently $\dot{\mu} = \dot{\bar{\mu}}$), as desired).

In the opposite case (i.e. $\phi < \phi^{*}$, though still $\mu \geq \mu^{*}$) the presence of the generally non-null second term in (60) might instead act in such a way as to oppositely mask the positive contribution given by the first term; thus possibly leading to a decreasing behavior for *MM* which might become lower than μ^{*}, and thus pushing toward singularities even in cases of desired end-effector motions totally evolving inside *DRW*.

Naturally enough, according to intuition, the occurrence of singularities becomes an unavoidable event whenever goal frame $< g >$, for some reasons, is located outside *DRW*.

Appendix C

Regarding the scalar condition (12), and related solution (13) under the assumed inequality $\mu^* \leq \mu < \mu_0$, it should be explicitly noted how the unique situations where it cannot be fulfilled actually occur only in correspondence of row vectors k exhibiting negligible square norms λ, that is, from a practical point of view, such that $\lambda < \lambda^*$, being λ^* the small regularization threshold used within (13). In such situation, however, a tendency of λ toward zero simply means a tendency of the non-zero vector p^T (non-zero since $\mu \geq \mu^*$) to become orthogonal to $R(J)$, as established by (9). This in turn implies a tendency of p^T itself to completely lie on the orthogonal complement $N(J)$; it is consequently clear how a suitable choice for both regularization thresholds ϕ^*, λ^* can actually be made so as to ensure, even for $\lambda < \lambda^*$, an increasing behaviour of μ via again the same mechanism of Appendix B, for $\phi \geq \phi^*$.

Methods and Algorithms for Sensor Data Fusion Aimed at Improving the Autonomy of a Mobile Robot

Andrea Bonci, Gianluca Ippoliti, Leopoldo Jetto, Tommaso Leo, and Sauro Longhi

Dipartimento di Ingegneria Informatica Gestionale e dell'Automazione
Università Politecnica delle Marche
Via Brecce Bianche, 60131 Ancona, Italy
<*a.bonci,g.ippoliti,l.jetto*>*@ee.univpm.it*, <*tommaso.leo,sauro.longhi*>*@univpm.it*
http://www.univpm.it

Abstract. A basic requirement for an autonomous mobile robot is to localize itself with respect to a given coordinate system. In this regard two different operating conditions exist: structured and unstructured environment. The relative methods and algorithms are strongly influenced by the *a priori* knowledge on the environment where the robot operates. If the environment is known, a proper multisensor system endowed with an efficient data fusion algorithm may provide a very accurate localization. In this chapter the localization problem is formulated in a stochastic setting and a Kalman filtering approach is proposed for the integration of odometric, gyroscope, sonar and video camera measures. If the environment is only partially known the localization algorithm needs a preliminary definition of a suitable environment map. Different probabilistic methods for sensory data fusion aimed at increasing the environment knowledge are proposed and discussed.

1 Introduction

To improve the performance of a mobile robot, a primary need is to increase its autonomy by enhancing the capability of localization with respect to the surrounding environment. This gives rise to the so-called Pose Estimation Problem and Map Building Problem. For their solution a growing interest in the study, development and analysis of many different kinds of sensory devices and perception systems can be recognized. In particular, research interests focused on multiple-sensor systems because of the limitations inherent any single sensory device, that can only supply a partial information on the environment, thus limiting the ability of the robot to localize itself. When a multi-sensor information is used, different kinds of observations are obtained. These observations are always affected by several kinds of uncertainties and are often partial, sparse and incomplete. This explains the great deal of research devoted to developing a methodology for an efficient integration of multiple-sensor information. The methods and algorithms proposed in the literature differ according to the *a priori* information on the environment, which may be almost known and static, or almost unknown and dynamic.

Recently, the Simultaneous Localization And Map building problem (SLAM problem) has been also deeply investigated for increasing the autonomy of navigation of mobile robots (see e.g. [68,24,62,69,70,22,71,18,23,35,36,16,3,25,30,48,54,55,67]). The idea of developing a mobile robot that can build a map of its environment while

B. Siciliano et al. (Eds.): Advances in Control of Articulated and Mobile Robots, STAR 10, pp. 191–222, 2004.
© Springer-Verlag Berlin Heidelberg 2004

simultaneously using that map to localize itself promises to allow these vehicle to operate autonomously for long period of time in unknown environments. Many contributions have been focused on the use of stochastic estimation techniques to build and maintain current estimates of vehicle position and of the environment map comprehending specific features location (frequently landmarks location). In particular, the Extended Kalman Filter (EKF) has been proposed as a tool for consistent fusion of the information acquired by the robot to yield estimates of vehicle and landmark locations by a recursive approach [22,25,68,24,71,35,36,3]. Another approach that has received considerable interest in the literature is based on a probabilistic method. For example in [70,62,69] an algorithm based on a rigorous statistical account of robot motion and perception is proposed for landmark based map acquisition and concurrent localization. Besides these approaches based on the rigorous mathematical models of the vehicle and sensing properties, different solutions have been proposed using a more qualitative knowledge of the nature of the environment [14,48,55]. The promising approach appears to be the one based on the use of stochastic estimation techniques, where an EKF is used for calculating the current position and orientation of the mobile robot which is subsequently fed to a map-building algorithm. To obtain an efficient integration of map building and localization, the acquired knowledge on the environment must be represented by parametric features with the associated uncertainty. The aim is to integrate in the same filtering algorithm the robot pose estimation and the environment features estimation. Moreover an adaptive algorithm is necessary to cope with the uncertainties on the environment and on the sensor readings.

Therefore two aspects are relevant for developing an efficient SLAM algorithm, the robot pose estimation and the environment features extraction. Both these aspects are considered in this chapter. Indoor environments are considered and 2D environment model is developed. Many real applications can be handled by this solution as for example in the emerging area of assistive technologies where powered wheelchairs can be used to strengthen the residual abilities of users with motor disabilities [58,32,12].

1.1   The Pose Estimation Problem

The pose estimation problem is to localize the robot with respect to an *a priori* known environment. Indoor environments are considered and 2D models are used where the known environment features are modeled by straight lines. Two different kinds of robot localization exist: relative and absolute. The first one is realized through the measures provided by sensors measuring variables internal to the vehicle (internal sensors). Typical internal sensors are optical incremental encoders which are fixed to the axis of the driving wheels or to the steering axis of the vehicle. At each sampling time the position is estimated on the basis of the encoder increments along the sampling interval. A drawback of this method is that the errors of each measure are summed up as movement proceeds. This heavily degrades the position and orientation estimates of the vehicle, in particular for long and winding trajectories [66]. In [10] practical methods are proposed to reduce odometry errors due to uncertainty about the effective wheelbase and unequal wheel diameter. Other typical internal sensors are gyroscopes and accelerometers which provide angular rate information and velocity rate information, respectively. The information provided by these inertial sensors must be integrated to obtain absolute estimates of orientation, position and velocity. Therefore, like for the odometers, even small errors in the individual measures may give rise to unbounded errors in the absolute measure.

Absolute localization is performed by processing the data provided by a proper set of external sensors measuring some parameters of the environment in which the vehicle is operating. A set of sonars is generally used as external sensory device. Sonars are fixed to the vehicle and measure the distance with respect to parts of the known environment [26,20,51,52,39,21,65,17]. The characterization of sonar measures and/or the rejection of unreliable sonar readings have been widely investigated [52,13,5,11]. Also a video camera can be used as external sensor. It is fixed to the vehicle and provides information on the characteristics features of the environment. Both sonars and video cameras are also widely utilized for the map building as required in the guidance of autonomous vehicles with obstacle avoidance in unknown environments [29,47,31,2,7].

The main drawback of absolute measures is their dependence on the characteristics of the environment. Possible changes of environmental parameters may give rise to erroneous interpretation of the measures provided by the localization algorithm. The actual trend is to exploit the complementary nature of internal and external sensors and to properly weight the relative data according to their reliability. For this purpose Kalman filtering techniques represent a powerful tool [21,17,64,46,34].

In this approach the internal and external sensors readings are combined together through an Extended Kalman Filter (EKF) providing on line estimates of robot position.

The use of Kalman filtering techniques requires to derive a stochastic state-space representation of the robot model and of the measure process. Formally this can be readily performed by applying the kinematic model of the robot and the available knowledge on measurement equipment. An interesting feature of the EKF here proposed is its capability of adaptively estimating the state and measurement noise covariance matrices.

1.2   The Map Building Problem

In this case the problem is to build a map (generally local) of the environment. External sensors can be also used for such a purpose [28,53,2]. The map building problem has been addressed by many researchers and over the years two basic approaches to environment representation have been developed: Grid-Based Modeling (GBM) and Feature-Based Modeling (FBM). In these approaches the environment is unknown and an accurate estimation of the robot pose is necessary. Range sensors are used for acquiring environment data that are the distance readings between the selected environment features and the robot. When dead-reckoning sensors are used for the pose estimation a poor environment model is generally obtained for long trajectories. This requires the fusion of all sensors readings in an efficient algorithm to simultaneously handle the map building and pose estimation as discussed in Section 5.

In the GBM approach the workspace of the robot is decomposed into square areas denoted as cells. In each cell a value, that corresponds to the level of certainty that an obstacle exists within the cell area, is stored (occupancy grid).

A characteristic of the structured environments is that objects tend to have straight borders. Indoors environments can be represented by a collection of line segments, representing the vertical surfaces of walls, doors, objects, etc. In the FBM approach, line segments or surfaces are used for modeling indoor environment and for improving the estimated position and orientation of the mobile robot (robot pose estimation) as recalled in the previous section [59,3]. Line features can be also detected in the occupancy grids as aligned cells of high probability of occupation.

The occupancy grid map is generally used for local path planning and reactive navigation; it is implemented by a variety of algorithms. Its main drawbacks are the difficulty in using grids to improve the robot pose estimation and the amount of computer memory needed for representing large environments. On the other hand, the FBM uses the parametric features for describing the boundaries of free-space in terms of lines or surfaces defined by a list of parameters (geometric primitives). This is useful for the local path planning and for the pose estimation. If sonar sensors are used for acquiring environment information, the uncertainties of sensor readings make unreliable the process of grouping (sonar) readings in geometric primitives; for example, multiple reflections can make sonar measurements erroneous for mapping corners in a square environments. In general, the integration with the readings of a video camera reduces the problems of grouping adjacent sensor measurements for obtaining more reliable environment features.

A method is here proposed for modeling the robot environment by extracting parametric straight line features and the associated uncertainty level both from the occupancy grid map and from the video data acquired by a CCD camera. The environment model is a 2D map which represents the 3D environment of the robot as a collection of line features estimating the boundaries of the environment.

1.3 Table of Contents

The possible integration of the pose estimation problem with the map building problem is preliminarly analyzed. In the following section the considered set of sensors will be presented. In Section 3 the algorithms developed for on line estimation of robot position will be analyzed and experimental results will be presented and discussed.

A multisensor fusion approach for improving the map-building capability of a mobile robot will be presented in Section 4. A modelling technique for indoor environments based on straight line features extraction from video data and sonar readings will be analyzed. The Hough Transform (HT) is considered for extracting straight lines from the occupancy grid map and from video data. In this section experimental results will be presented and analyzed.

These algorithms give good performance for the robot pose estimation and for the environment features extraction in a rigorous mathematical unified framework. Therefore these results are promising for the solution of the SLAM problem that will be discussed in Section 5.

2   The Sensory Equipment

The methods and algorithms developed in this chapter refer to the vehicle of Fig. 1. It is an unicycle-like mobile robot with two driving wheels, mounted on the left and right sides of the robot, with their common axis passing through the geometrical center of the robot (see Fig. 1). Localization of this mobile robot in a two-dimensional space requires three free coordinates: coordinates x and y of the midpoint between the two driving wheels and the angle θ between the main axis of the robot and the x-direction. The kinematic model of the unicycle robot is described by the following equations:

$$\dot{x}(t) = \nu(t)\cos\theta(t) \tag{1}$$
$$\dot{y}(t) = \nu(t)\sin\theta(t) \tag{2}$$
$$\dot{\theta}(t) = \omega(t), \tag{3}$$

where $\nu(t)$ and $\omega(t)$ are, respectively, the displacement and angular velocities of the robot.

2.1   Odometric Measures

The encoders placed on the driving wheels provide a measure of the incremental angles over a sampling period $\Delta t_k := t_{k+1} - t_k$. The odometric measures are used to obtain an estimate of the linear and angular velocities $\bar{\nu}(t_k)$ and $\bar{\omega}(t_k)$, respectively, which are assumed to be constant over the sampling period. Numerical integration of (1) and (2) based on $\bar{\nu}(t_k)$ and $\bar{\omega}(t_k)$ provides an estimate of the position and orientation increments over each sampling period of the unicycle robot. Such processing is generally performed by an odometric device connected with the low level controller of the robot (imposing the desired $\nu(t_k)$ and $\omega(t_k)$).

The encoders incremental errors heavily affect the estimate of the orientation θ; this limits their applicability to short trajectories only. An analysis of the accuracy of the estimation procedure implemented by an odometric equipment has been developed in [66].

2.2   Fiber Optic Gyroscope Measures

The accuracy of the robot pose estimation can be greatly improved by the use of the Fiber Optic Gyroscope (FOG), that provides very reliable measures of the orientation θ.

The operation principle of a Fiber Optic Gyroscope (FOG) is based on the Sagnac effect. The FOG is made of a fiber optic loop, fiber optic components, a

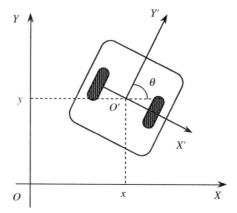

Fig. 1. Scheme of the unicycle robot.

photo-detector and a semiconductor laser. The phase difference of two light beams traveling in opposite directions around the fiber optic loop is proportional to the rate of rotation of the loop. The rate information is internally integrated to provide the absolute measurements of orientation. A FOG does not require frequent maintenance and has a lifetime longer than the conventional mechanical gyroscopes. The drift is also low. A careful analysis on the accuracy of this internal sensor has been developed in [57].

2.3   Sonar Measures

The distance readings by sonar sensors are related to the indoor environment model and to the configuration of the mobile robot.

Consider a planar distribution of n_s sonar sensors. Denote with x'_i, y'_i, θ'_i the position of the i-th sonar, $i = 1, 2, \ldots, n_s$, referred to the coordinate system (O', X', Y') fixed to the mobile robot, as reported in Fig. 2.

The position x_i, y_i, θ_i at the sampling time t_k of the i-th sonar referred to the inertial coordinate system (O, X, Y) have the following form:

$$x_i(t_k) = x(t_k) + x'_i \sin\theta(t_k) + y'_i \cos\theta(t_k), \tag{4}$$
$$y_i(t_k) = y(t_k) - x'_i \cos\theta(t_k) + y'_i \sin\theta(t_k), \tag{5}$$
$$\theta_i(t_k) = \theta(t_k) + \theta_i. \tag{6}$$

The walls and the obstacles in an indoor environment are represented by a proper set of planes orthogonal to the plane XY of the inertial coordinate system. Each plane $P^j, j = 1, 2, \ldots, n_p$ (where n_p is the number of planes which describe the indoor environment) is represented by the triplet P^j_r, P^j_n, P^j_ν, where P^j_r is the normal distance of the plane from the origin O, P^j_n is the angle between the normal line to the plane and the x-direction and P^j_ν is a binary variable, $P^j_\nu \in \{-1, 1\}$, which defines

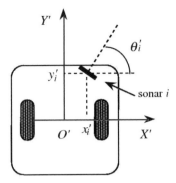

Fig. 2. Sonar displacement.

the face of the plane reflecting the sonar beam. In such a notation, the expectation $d_i^j(t_k)$ for the present distance of the sonar i from the plane P^j has the following expression (see Fig. 3):

$$d_i^j(t_k) = P_\nu^j(P_r^j - x_i(t_k)\cos P_n^j - y_i(t_k)\sin P_n^j),\tag{7}$$

if the $P_n^j \in [\theta_i(t_k) - \delta/2, \theta_i(t_k) + \delta/2]$, where δ is the beamwidth of the sonar sensor. The vector composed of geometric parameters $P_r^j, P_n^j, P_\nu^j, j = 1, 2, \ldots, n_p$, is denoted by Π.

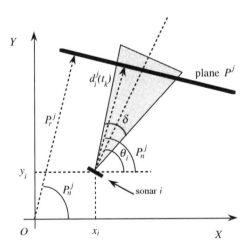

Fig. 3. Sonar measure.

To simplify the position estimation algorithm without appreciable reduction of accuracy, the sonar echoes traveling along the cone edges have been omitted. In fact, the measures along the cone edges require an *a priori* model of the environment including the different roughness of the walls and are less accurate of the distance measures given by (7).

2.4 Video Camera Measures

A video camera and related image processing procedures can be used for extracting environment features that are related to the indoor environment model and to the robot configuration.

Consider a CCD video camera installed on a mobile robot. For the image formation, reference is made to the pinhole model [4]. This is a simple video camera model, which does not take into account some linear and nonlinear distortions phenomena in the image formation process [60,33]. The main linear distortions are relative to the image center displacement [50] and to the scale difference [63,38]. Taking these phenomena into account, the CCD calibration equations defining the relationship between the metric coordinates of a point p_w with its pixel coordinates have the following form:

$$
\begin{aligned}
u &= u_0 + \frac{s_u x}{d_x} = \frac{s_u x + c_x}{d_x}, \\
v &= v_0 + \frac{y}{d_y} = \frac{y + c_y}{d_y},
\end{aligned}
\tag{8}
$$

where s_u is the horizontal scale factor [63], $d_x := S_{d_x}/N_{d_x}$ is the center-to-center distance between adjacent CCD sensor elements in the x direction (scan line), $d_y := S_{d_y}/N_{d_y}$ is the center-to-center distance between adjacent CCD sensor elements in the y direction, S_{d_x}, S_{d_y} are the CCD sizes, N_{d_x}, N_{d_y} are the numbers of CCD elements and c_x, c_y are the row and column indices of the center of the digital image. Therefore the complete set of camera parameters that must be estimated are the intrinsic parameters f, s_u, c_x, c_y, where f is the focal length, and the set of extrinsic parameters that determine the position and orientation of the video camera referred to the environment frame (see [7]). This model is appropriate for modern solid-state cameras, especially in the context of mobile robotics [4].

Different camera calibration techniques are proposed into the literature (see e.g. [50]). The algorithm recently proposed in [38] is here used for the estimation of the video camera parameters.

To reduce the computation efforts, the "visible space" is introduced. In the pinhole model, the viewing frustum of the video camera is the projection of the image plane corners from the pinhole, that is located one focal length behind the image plane [45]. Pointing the camera down in the forward direction of the robot, the "visible space" on the ground plane is defined by the projection of the frustum vertices on the floor plane (see Fig. 4).

Moreover, for improving the detection of environment's features, the Hough Transform (HT) is used [40,27,56].

In the HT the straight line equation, is expressed by:

$$
\rho = u \cos \phi + v \sin \phi,
\tag{9}
$$

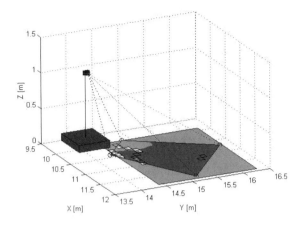

Fig. 4. Visible space of the CCD camera installed on the robot.

where ρ is the distance between the straight line and the origin of the coordinate system in the image plane, ϕ is the orientation of the line and u,v are the coordinates of whatever edge point belonging to the line. The HT requires an accumulator array $H(\rho,\phi)$, called Hough space, to represent the possible values of (ρ,ϕ); it is generally approximated by a discrete array. The edge points (u,v) are detected by means of an orthogonal differential operator [37] (e.g., the Sobel operator [61]); for each detected edge point the parameters (ρ,ϕ) are estimated and quantized, and the accumulator array is increased accordingly. After this preliminary edge points processing, the accumulator array is searched for peaks. The peaks identify the parameters of the highest probability lines. In the standard HT the accumulator is increased by the same quantity for each edge point, assuming that each of them contribute equally to the line features.

3   Estimation of Robot Location by Kalman Filter

In this section an EKF is proposed for the on line estimation of robot location through the fusion of internal and external sensors measures. The environment is assumed to be *a priori* known.

3.1   Extended Kalman Filter

Denote with T the sampling period and let $X(kT) := \begin{bmatrix} x(kT) & y(kT) & \theta(kT) \end{bmatrix}^T$ and $Z(kT)$ be the robot state and the measurement vector respectively at time kT. Vectors $Z(kT)$ and $X(kT)$ can be related by a nonlinear measure equation of the kind

$$Z(kT) = G(X(kT), \Pi) + v(kT), \tag{10}$$

where $G(X(kT), \Pi)$ is a nonlinear function of the state and of the geometric parameter vector Π and $v(kT)$ is a white noise sequence $\sim N(0, R(kT))$. The dimension
p_k of $Z(kT)$ is not constant, depending on the number of sensory measures that are
actually used at each time.

Let $U(kT) := [v(kT), \omega(kT)]$ be the robot control input at time kT and assume
$U(t) = U(kT)$ for $t \in [kT, (k+1)T)$.

To obtain an EKF with an effective state prediction equation in a simple form,
model (1)–(3) has been linearized about the current state estimate $\hat{X}(kT, kT)$ and
the control input $U((k-1)T)$ applied until the linearization instant. Subsequent
discretization with period T of the linearized model gives the following EKF (where
explicit dependence on T has been dropped for simplicity of notation),

$$\hat{X}(k+1, k) = \hat{X}(k, k) + L(k)\hat{X}(k, k) \tag{11}$$

$$P(k+1, k) = A_d(k)P(k, k)A_d^T(k) + Q_d(k) \tag{12}$$

$$K(k+1) = P(k+1, k)C^T(k+1) \cdot$$
$$[C(k+1)P(k+1, k)C^T(k+1) + R(k+1)]^{-1} \tag{13}$$

$$\hat{X}(k+1, k+1) = \hat{X}(k+1, k) + K(k+1) \cdot$$
$$[Z(k+1) - G(\hat{X}(k+1, k), \Pi)] \tag{14}$$

$$P(k+1, k+1) = [I - K(k+1)C(k+1)] \cdot$$
$$P(k+1, k), \tag{15}$$

where:

$$L(k) := \begin{bmatrix} T\cos\theta(k) & -0.5v(k-1)T^2\sin\theta(k) \\ T\sin\theta(k) & 0.5v(k-1)T^2\cos\theta(k) \\ 0 & T \end{bmatrix}, \tag{16}$$

$$A_d(k) := \begin{bmatrix} 1 & 0 & -v(k-1)\sin\theta(k) \\ 0 & 1 & v(k-1)\cos\theta(k) \\ 0 & 0 & 1 \end{bmatrix} \tag{17}$$

$$Q_d(k) := \sigma_\eta^2(k)\bar{Q}(k) \tag{18}$$

$$\bar{Q}(k) := \begin{bmatrix} T + v(k-1)^2\frac{T^3}{3}\sin^2\theta(k) \\ -v(k-1)^2\frac{T^3}{3}\cos\theta(k)\sin\theta(k) \\ -v(k-1)\frac{T^2}{2}\sin\theta(k) \end{bmatrix}$$

$$\begin{matrix} -v(k-1)^2\frac{T^3}{3}\cos\theta(k)\sin\theta(k) & -v(k-1)\frac{T^2}{2}\sin\theta(k) \\ T + v(k-1)^2\frac{T^3}{3}\cos^2\theta(k) & v(k-1)\frac{T^2}{2}\cos\theta(k) \\ v(k-1)\frac{T^2}{2}\cos\theta(k) & T \end{matrix} \Bigg], \tag{19}$$

and $C(k)$ is the $(3 \times p_k)$ matrix obtained by the linearization of the measure equation
(10).

The form of $Q_d(k)$ expressed by (18) derives by the hypothesis that model
(1)–(3) describes the true dynamics of the three state variables with nearly the same

degree of approximation and with independent errors. The structure of $R(k)$ depends on the particular sensor equipment used.

In the next sections, the above general structure of the EKF will be particularized according to the sensor equipment used.

3.2   Adaptive Estimation of $Q_d(k)$ and $R(k)$

The EKF can be implemented when estimates of $Q_d(k)$ and $R(k)$ are available.

A considerable amount of research has been performed on the adaptive Kalman filtering (see [1,41,15] and references therein), but in practice it is often necessary to redesign the adaptive filtering scheme according to the particular characteristics of the problem faced. The adaptive procedure here proposed refers to matrices $Q_d(k)$ of the form (18) and $R(k) = \text{diag}\,[\sigma_{v,i}^2,\; i = 1, \cdots, p_k]$.

The following nearly time-invariance assumption is here made: the parameters $\sigma_{v,i}^2(k),\; i = 1, \dots, p_k$, and $\sigma_\eta^2(k)$ are nearly constant over $n_v \geq 2$ and $n_\eta \geq 2$ samples respectively.

Define $\gamma_i(k + 1) = z_i(k + 1) - G_i(\hat{X}(k + 1, k), \Pi)$, where $z_i(k + 1)$ and $G_i(\hat{X}(k + 1, k), \Pi)$ are the i-th component of $Z(k + 1)$ and $G(\hat{X}(k + 1, k), \Pi)$, respectively. In analogy with the linear case, residuals $\gamma_i(k + 1), i = 1, \dots, p_k$, are named the innovation process samples and are assumed to be well described by a white sequence $\sim N(0, s_i(k + 1))$, where $s_i(k + 1), i = 1, \dots, p_k$ can be expressed as

$$
\begin{aligned}
s_i(k + 1) &= C_i(k + 1)P(k + 1, k)C_i^T(k + 1) + \sigma_{v,i}^2(k + 1) \\
&= C_i(k + 1)[A_d(k)P(k, k)A_d^T(k) + \sigma_\eta^2(k)\bar{Q}(k)]\,\cdot \\
&\qquad\qquad\qquad C_i^T(k + 1) + \sigma_{v,i}^2(k + 1), \qquad (20)
\end{aligned}
$$

where $C_i(\cdot)$ is the i-th row of $C(\cdot)$.

This simplifying assumption is valid as long as discretization and linearization of (1)–(3) and (10) is accurate and allows the extension of the methods of the adaptive filtering theory developed for the linear case.

The two above assumptions will allow defining a simple and efficient estimation algorithm based on the condition of consistency, at each step, between the observed innovation process samples $\gamma_i(k + 1), i = 1, \dots, p_k$ and their predicted statistics $E\{\gamma_i^2(k + 1)\} = s_i(k + 1)$. Imposing such a condition, one-stage estimates $\hat{\sigma}_\eta^2(k)$ and $\hat{\sigma}_{v,i}^2(k+1), i = 1, \dots, p_k$, of $\sigma_\eta^2(k)$ and $\sigma_{v,i}^2(k+1), i = 1, \dots, p_k$, respectively, are obtained at each step. To increase their statistical significance, the one-stage estimates $\hat{\sigma}_\eta^2(k)$ and $\hat{\sigma}_{v,i}^2(k + 1), i = 1, \dots, p_k$, are averaged obtaining the relative smoothed versions $\bar{\hat{\sigma}}_\eta^2(k)$ and $\bar{\hat{\sigma}}_{v,i}^2(k + 1), i = 1, \dots, p_k$.

After proper calculations (see [42] for details), the following recursive form of estimates $\bar{\hat{\sigma}}_\eta^2(k)$ and $\bar{\hat{\sigma}}_{v,i}^2(k + 1),\; i = 1, \dots, p_k$, is found

$$\bar{\bar{\sigma}}_\eta^2(k) = \bar{\bar{\sigma}}_\eta^2(k-1) +$$

$$\frac{1}{(l_\eta + 1)p_k} \left[\sum_{i=1}^{p_k} \left(\hat{\sigma}_{\eta,i}^2(k) - \hat{\sigma}_{\eta,i}^2(k - (l_\eta + 1)) \right) \right] \tag{21}$$

$$\bar{\bar{\sigma}}_{v,i}^2(k+1) = \bar{\bar{\sigma}}_{v,i}^2(k) + \frac{1}{l_v + 1}(\hat{\sigma}^2(k+1) - \hat{\sigma}^2(k - l_v)), \tag{22}$$

where:

- $\hat{\sigma}_{\eta,i}^2(k) = \max \left\{ (C_i(k+1)\bar{Q}(k)C_i^T(k+1))^{-1}[\gamma_i(k+1)^2 - \right.$
  $$C_i(k+1)A_d(k)P(k,k)A_d^T(k)C_i^T(k+1) - \bar{\bar{\sigma}}_{v,i}^2(k+1)], 0 \right\}$$
- $\hat{\sigma}_{v,i}^2(k+1) = \max \left\{ \gamma_i^2(k+1) - [C_i(k+1)A_d(k)P(k,k) \cdot \right.$
  $$\left. A_d^T(k)C_i^T(k+1) + C_i(k+1)\bar{\bar{\sigma}}_{\eta,i}^2(k)\bar{Q}(k)C_i^T(k+1)], 0 \right\},$$
- l_η and l_v are the number of one-stage estimates $\bar{\bar{\sigma}}_\eta^2(k)$ and $\bar{\bar{\sigma}}_{v,i}^2(k+1)$ respectively, yielding the smoothed estimates.

Parameters l_η and l_v of estimators (21) and (22) are chosen on the basis of two antagonist considerations: low values would produce noise estimators which are not statistically significant; large values would produce estimators which are scarcely sensitive to possible rapid fluctuations of the true $\sigma_\eta^2(k)$ and $\sigma_{v,i}^2(k)$, $i = 1, \ldots, p_k$. During filter initialization, the starting values $(\hat{\sigma}_\eta^0)^2$ and $(\hat{\sigma}_{v,i}^0)^2$, $i = 1, \ldots, p_k$, of $\hat{\sigma}_\eta^2(k)$ and $\hat{\sigma}_{v,i}^2(k)$ respectively, must be chosen on the basis of the *a priori* available information. In case of lack of such information, a large value of $P(0,0)$ is useful to prevent divergence.

With some formal variants, the above procedure can be extended to the case where also the covariance matrix of the measurement noise is $R(k) = \sigma_v^2(k)\bar{R}(k)$, $R(k)$ being a known matrix.

A recent alternative approach based on a fuzzy adaptation mechanism has been proposed in [43].

3.3   Sensors Readings Selection

To reduce the probability of an inadequate interpretation of erroneous sensor data, a method is proposed here to deal with the undesired interferences produced by the presence of unknown obstacles on the environment or by incertitude on the sensor readings. Notice that for the problem here handled both the above events are equally distributed. A simple and efficient way to perform this preliminary measure selection is to compare the actual sensor readings with their expected values. Measures are discharged if the difference exceeds an adaptively time-varying threshold. This is here done in the following way: at each step, for each measure $z_i(\cdot)$ of an external sensor, the residual $\gamma_i(k+1) = z_i(k+1) - G_i(\hat{X}(k+1,k), \Pi)$, represents the difference between the actual sensor measure $z_i(k+1)$ and its expected value

$G_i(\hat{X}(k+1,k), \Pi)$ which is computed on the basis of the estimated robot location and on the *a priori* knowledge of the environment. As $\gamma_i \sim N(0, s_i(k+1))$, the current value $z_i(k+1)$ is accepted if $|\gamma_i(k+1)| \leq 2\sqrt{s_i(k+1)}$. Namely, the variable threshold is chosen as two times the standard deviation of the innovation process.

3.4   Pose Estimation by Fusion of Odometric and Inertial Measures

If only internal sensors are used, the measure Eq. (10) reduces to

$$Z((k+1)T) = X((k+1)T) + V(kT), \tag{23}$$

where $Z(kT) = [z_1(kT), z_2(kT), z_3(kT)]^T$ and $V(kT) = [v_1(kT), v_2(kT), v_3(kT)]^T$ is a white sequence $\sim N(0, R(kT))$. The elements of $Z(kT)$ are: $z_1((k+1)T) \equiv x_d((k+1)T)$, $z_2((k+1)T) \equiv y_d((k+1)T)$, $z_3((k+1)T) \equiv \theta_g((k+1)T)$, where $x_d((k+1)T)$ and $y_d((k+1)T)$ are computed through classical odometric algorithms exploiting the angular measure $\theta_g((k+1)T)$ provided by the FOG.

The covariance matrix $R(kT)$ of $V(kT)$ has the following structure:

$$R(kT) = \text{block diag} \left[\sigma_o^2(kT)\overline{R}(kT), \sigma_g^2\right] \tag{24}$$

where the scalar $\sigma_o^2(kT)$ is the measurement noise variance depending on the odometers; $\overline{R}(kT)$ is a (2×2) matrix that can be composed through the equations of the used odometric algorithm; σ_g^2 is the constant variance of the noise $v_3(kT)$ affecting $\theta_g(kT)$. As the measures provided by the FOG are much more reliable than the other ones, one has $\sigma_g^2 \ll \sigma_o^2$ and a nearly singular filtering problem is obtained.

In this case a lower order non-singular EKF can be derived assuming that the original $R(kT)$ is actually singular [1].

The experimental tests performed with this set of sensors are discussed beneath. A commercial powered wheelchair TGR Explorer has been used. This vehicle has been developed to be used in the emerging area of assistive technologies where powered wheelchairs can be used to strengthen the residual abilities of users with motor disabilities. A control module in the guidance system is developed for translating the commands generated by the navigation module or by the user in the driving commands for the actuators of the wheelchair (see [32]).

The implementation of the navigation system for this mobile base was performed on a PC 486DX2 with PC-104 bus installed on the rear side of the wheelchair (see Fig. 5). The PC installed on the wheelchair also manages the sensory system and the connection with the user interface. The sensory system is based on FOG sensor and odometric sensors that allow the estimation of the mobile base position with respect to a starting reference configuration. The odometric system has been simply carried out by two incremental optical encoders aligned with the axes of the driving wheels. The gyroscopic measures on the absolute orientation have been collected in a digital form by a serial port on the computer. The fiber optic gyroscope HITACHI mod. HOFG-1 was used for measuring the angle θ of the mobile robot.

The EKF has been implemented on a MS Windows PC by the development environment described in [9]. In this development system, the planned trajectory

Fig. 5. The PC-104 bus installed on the wheelchair with data acquisition system for the FOG sensor and the incremental encoders.

has been computed considering the non-holonomic and environment constraints according to the algorithm proposed in [19]. The system is connected directly with the low level robot controller by standard serial protocol RS232. All the experiments have been performed on closed trajectories making the robot track relatively long. A sample of the performed experimental tests is shown in Fig. 6. Part (a) of this figure shows the estimated trajectory with the localization algorithm based only on odometric measures. A long trajectory of 108 meters has been considered to verify the limitations intrinsic to the use of odometric measures. The plot clearly evidences the unreliability of the estimated trajectory. Part (b) shows the same test with the localization algorithm based on both odometric and inertial measures. The plot clearly shows the improvement introduced: at the end of the test trajectory the error on the pose estimation is of 16 cm.

3.5   Pose Estimation by Fusion of Odometric and Sonar Measures

In this case the measure vector $Z(kT)$ is composed of two subvectors $Z_1(kT) = [\, z_1(kT) \;\; z_2(kT) \;\; z_3(kT)\,]^T$ and $Z_2(kT) = [\, z_4(kT) \;\; z_5(kT) \;\; ... \;\; z_{3+n_s}(kT)\,]^T$, where $z_1((k+1)T) = x_d((k+1)T)$, $z_2((k+1)T) = y_d((k+1)T)$, $z_3((k+1)T) = \theta_d((k+1)T)$ are the measures provided by the odometric device, and $z_{3+i}((k+1)T) = d_i^j((k+1)T) + v_{3+i}((k+1)T)$, $i = 1, 2, \ldots, n_s$, $j \in [1, n_p]$, with $d_i^j((k+1)T)$ given by (7), is the distance measure provided by the i-th sonar sensor from the P^j plane with $j \in [1, n_p]$. The environment map provides the information needed to detect which is the plane P^j in front of the i-th sonar.

By definition of the measurement vector one has that the output function $G(X((k+1)T), \Pi)$ has the following form:

$$G(X((k+1)T), \Pi) = [\, x((k+1)T) \;\; y((k+1)T) \;\; \theta((k+1)T)$$

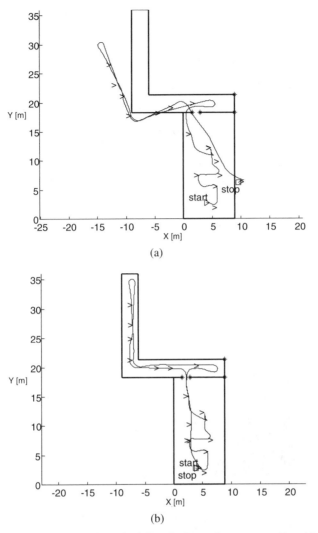

Fig. 6. Pose estimation by fusion of FOG and odometric measures. Part (a): estimated trajectory with localization algorithm based on odometric measures only. Part (b): estimated trajectory with localization algorithm based on FOG and odometric measures.

$$d_1^{j_1}((k+1)T) \; d_2^{j_2}((k+1)T) \; \cdots \; d_{\bar{p}_{k+1}}^{j_{\bar{p}_{k+1}}}((k+1)T) \Big]^T$$
$$j_1, j_2, \ldots, j_{\bar{p}_{k+1}} \in [1, n_p], \tag{25}$$

where $\bar{p}_k := p_k - 3$. The number p_k of measures may vary from the minimum value 3 to the maximum value $n_s + 3$, where n_s is the number of sonar sensors.

Matrix $C(k)$ has the following form

$$C(k) := [\, C_1(k)^T \quad C_2(k)^T \quad \cdots \quad C_{p_k}(k)^T \,]^T, \tag{26}$$

where

$$[\, C_1(k)^T \quad C_2(k)^T \quad C_3(k)^T \,]^T = I_3,$$
$$C_{i+3}(k) = P_\nu^j \, (-\cos P_n^j - \sin P_n^j$$
$$+ x_i' \cos(\theta(k) - P_n^j) - y_i' \sin(\theta(k) - P_n^j))$$
$$i = 1, 2, \ldots, \bar{p}_k, \quad \bar{p}_k \le n_s, \quad j \in [1, n_p]. \tag{27}$$

The measurement noise covariance matrix $R(k)$ has the following structure: $R(k) =$ block diag $[R_1(k), \ R_2(k)]$. The block $R_1(k)$ is the (3×3) matrix according to the used odometric algorithm. The block $R_2(k)$ is a $((p_k-3 \times p_k-3))$ matrix representing the covariance matrix of the independent errors effecting the sonar measures.

A sample of the experimental tests performed with this set of sensors is reported beneath. The experimental tests have been carried out on the LabMate mobile base in an indoor environment with different geometries. This mobile robot is realized with two driving wheels, as reported in Fig. 1, and the odometric data are the incremental measures that, at each sampling interval, are provided by the encoders attached to the right and left wheels of the robot. These measures are directly captured by the low level controller of the mobile base. The sonar measures have been acquired by the standard proximity system of the LabMate base composed by a set of nine Polaroid sonar sensors. A picture of LabMate system with the sonar sensors placement is reported in Fig. 7. A preliminary reduction of crosstalk has been obtained by a proper

Fig. 7. Indoor environment with the LabMate mobile vehicle.

distribution on the orientations of the sonar sensors. A significant reduction of the wrong readings produced by unknown obstacles has been also realized following the procedure described in Section 3.3.

The localization algorithm has been tested with relatively long trajectories in an indoor environment represented by a suitable set of planes orthogonal to the plane XY of the inertial system.

Figure 8 illustrates the results of such an experiment. Part (a) of this figure represents the trajectory with localization deduced by odometric measures only: the pose estimation at the end of the considered trajectory is completely wrong and the robot crash into the wall. In order to test the limitation of the odometric measures, the planned trajectory is composed by a large set of orientation changes. The black path is the actual trajectory with only odometric measures. In this case at the end of the test the robot is out of the planned trajectory. Part (b) shows the same test with localization based on the Adaptive Extended Kalman Filter (AEKF) described in Section 3.2 and fed by odometric and sonar measures. The error on the pose estimation at the end of the planned trajectory is of 1.5 cm and the robot is able to go through the door. The values $l_\eta = l_v = 2$ have been chosen. The plot clearly evidences the improvement introduced by the adaptation mechanism.

3.6   Pose Estimation by Odometric and Video Camera Measures

As mentioned previously, a video camera can be used for identifying features of the environment. Detection of vertical straight lines features by HT has been considered. Each line is characterized by a pair of parameters (ρ_i, ϕ_i) where ρ_i is the distance between the line and the origin and ϕ_i specifies the orientation of the line. In this case, for each detected vertical straight line, a pair of measures (ρ_i, ϕ_i), depending on the state $\widehat{X}(kT)$, is produced. The output function $G(X((k+1)T), \Pi)$ has the following form:

$$G(X((k+1)T), \Pi) = [\, x((k+1)T)\ y((k+1)T)\ \theta((k+1)T)$$
$$\rho_1((k+1)T)\ \phi_2((k+1)T)\ \cdots\ \rho_{\bar{\bar{p}}_k}((k+1)T)\ \phi_{\bar{\bar{p}}_k}((k+1)T)\,]^T, \quad (28)$$

and the number of measures is $p_k = 3 + 2\bar{\bar{p}}_k$, where $\bar{\bar{p}}_k$ is the number of line features detected at time $(k+1)T$.

In this case the preliminary sensor reading selection described in Section 3.3 has been applied for a better exploitation of the measures which are related to the *a priori* knowledge of the environment.

The experimental tests have been performed in an indoor environment with different geometries. The same LabMate mobile base of Section 3.5 has been used; therefore, as for the odometers, the same considerations of that section hold. The video camera measures have been collected by a low cost CCD web-camera Philips PCVC 675K installed in front of the vehicle.

The localization algorithm has been tested over relatively long trajectories in an indoor environment represented by a suitable set of planes orthogonal to the plane XY of the inertial system.

The planned trajectory of Fig. 9 from the start configuration S to the goal configuration G is composed by a large set of orientation changes. If the localization

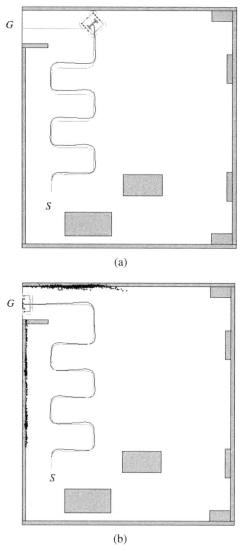

Fig. 8. Dots path is the planned trajectory from the start configuration S to the goal configuration G, the dark path is the realized trajectory: (a) localization with only odometric measures; (b) localization with the AEKF where the gray dots are the actually used sonar measures.

is obtained only through odometric measures, the end trajectory errors are 31.3 and 94.8 cm along the X and Y directions respectively. Introducing the video camera measures, a significant performance improvement has been obtained and the end trajectory error is of 8.9 cm. Figure 9 shows some samples of the images acquired along the considered trajectory and highlights the environment features used for the video camera readings.

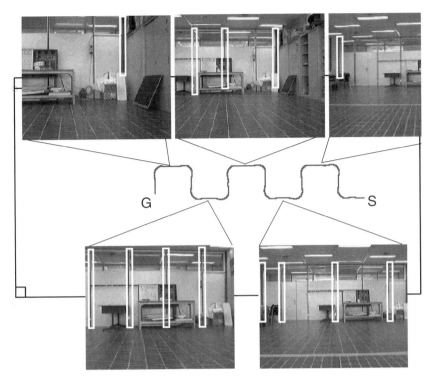

Fig. 9. Pose estimation by fusion of odometric and video camera measures.

4   Ultrasonic and Video Data Fusion for Map Building

The pose estimation described in the previous section assumes the *a priori* knowledge of the environment where the robot moves. Unfortunately, this is not the most frequent case. Therefore map building from sensory information collected by the robot itself has to be reliably performed. This section presents results in this respect following a multiple sensor approach. A structured environment has been assumed, and as clarified in the previous section, it is characterized by walls, doors, objects, etc. that are represented by straight lines. This section proposes a FBM of the environment in which straight line segments are used for modeling indoor environment and for improving the pose estimation of a mobile robot. This model keeps a selection of sensor readings produced by external sensors like sonar(s) and video camera(s).

The problem of line feature extraction is faced for both kinds of sensory data representations: occupancy grid, that is computed by probabilistic aggregation of sonar readings, and video data. The initial state of the occupancy grid is completely unknown because an *a priori* model of the environment is not provided. During the robot navigation the sonar readings are integrated into the occupancy grid and, at fixed time intervals, images of a part of the environment floor are acquired.

The obstacle borders on the floor have a geometry which can be generally described by straight lines. The proposed updating process makes use of a probabilistic
approach to the Hough Transform (HT) [40,37] for extracting line features and the
associated certainty values both from occupancy grid and from video data. Matching between the lines extracted from the occupancy grid and from video data is
performed for obstacles belonging to the part of the floor visible from the CCD
camera (here called "visible space" as in Section 2.4). The proposed matching algorithm is based on the combination of the lines probability encoded in both the
Hough accumulators.

4.1   Line Detection by Video Camera

The evaluation of the uncertainty of straight lines extracted from digital data is
relevant for building up the model.

Each line feature on the floor plane representing the i-th "wall" or "obstacle" of
the environment is represented by a vector defined as

$$w_i := [\Theta_i^T \ P(\Theta_i)]^T \tag{29}$$

where $\Theta_i = [\rho_i \quad \phi_i]^T$ is a vector representing the i-th line feature by means of
its polar coordinates: the orientation angle ϕ_i of the i-th line and the distance ρ_i
between the origin of the reference frame and the i-th line. $P(\Theta_i)$ is the probability
of existence of a line feature having parameters ρ_i and ϕ_i. In a recent contribution,
an algorithm has been introduced   [44] for updating the accumulator of the HT
depending from the uncertainty of each edge point. This algorithm makes use of
image noise, edge orientation estimation and parametric line representation, for
computing the variance σ_ρ^2, σ_ϕ^2 of the estimated line parameters $\hat{\phi}$ and $\hat{\rho}$ for each
edge point. The line parameters uncertainty are used for evaluating the joint density
function $p(\hat{\Theta}_C | \Theta_C)$, that is the likelihood of the all possible quantized values $\Theta_C =
[\rho \ \phi]^T$, given the observed line parameters $\hat{\Theta}_C = [\hat{\rho} \ \hat{\phi}]^T$. The assumption is
made that the variable $\hat{\Theta}_C$ is normally distributed as $\hat{\Theta}_C \sim N(\Theta_C, \Sigma_{\hat{\Theta}_C})$, where
$\Sigma_{\hat{\Theta}_C}$ is the covariance matrix of $\hat{\Theta}_C$

$$\Sigma_{\hat{\Theta}_C} = \begin{bmatrix} \sigma_\phi^2 & \sigma_{\rho\phi} \\ \sigma_{\rho\phi} & \sigma_\rho^2 \end{bmatrix}. \tag{30}$$

Under this assumption, $\hat{\Theta}_C$ has the following bivariate normal distribution

$$p(\hat{\Theta}_C | \Theta_C) = \frac{1}{2\pi} |\Sigma_{\hat{\Theta}_C}|^{-\frac{1}{2}} \exp\left(-\frac{1}{2}(\hat{\Theta}_C - \Theta_C)^T \Sigma_{\hat{\Theta}_C}^{-1} (\hat{\Theta}_C - \Theta_C)\right). \tag{31}$$

The Hough accumulator is incremented by the log $(p(\hat{\Theta}_C | \Theta_C))$ at each edge
point. The covariance matrix is singular $(|\Sigma_{\hat{\Theta}_C}| = 0)$, and thus $\hat{\Theta}_C$ is a singular or
degenerate bivariate normal distribution. This means that the probability density for
$\hat{\Theta}_C$ is always concentrated in a subspace whose dimension is smaller than that of the
space generated by $\hat{\Theta}_C$; hence the probability density distribution $p(\hat{\Theta}_C | \Theta_C)$ cannot

be directly computed. According to the properties of the bivariate joint distribution recalled in [49], the line parameters distribution can be described as

$$p(\hat{\Theta}_C|\Theta_C) = p(0,\hat{\phi}) = \frac{1}{\sqrt{2\pi}\sigma_\phi} \exp\left(-\frac{1}{2}\left(\frac{(\hat{\phi}-\phi)^2}{\sigma_\phi^2}\right)\right). \tag{32}$$

This means that the bivariate joint distribution of the two correlated random variables $\hat{\theta}$ and $\hat{\rho}$ can be computed in a simple way as the normal distribution (32) of only one of the two random variables. This variable is assumed to be $\hat{\theta}$. Therefore the probability that an edge point (x,y) belongs to the line whose parameters are Θ_C, given the observation $\hat{\theta}$, is simply obtained by integrating (32) as follows:

$$P(\hat{\Theta}_C|\Theta_C) = \frac{1}{\sqrt{2\pi}\sigma_\phi} \int_{\hat{\phi}-\frac{\triangle\phi}{2}}^{\hat{\phi}+\frac{\triangle\phi}{2}} \exp\left(-\frac{1}{2}\left[\frac{(\hat{\phi}-\phi)^2}{\sigma_\phi^2}\right]\right) d\phi, \tag{33}$$

where $\triangle\phi$ is the quantization step of ϕ in the Hough accumulator.

This result shows that the probability of a line does not depend on $\hat{\Theta}_C$, but only on the line orientation estimate $\hat{\phi}$. The computational effort needed for computing $P(\hat{\Theta}_C|\Theta_C)$ is therefore significantly reduced.

To evaluate the probability value of a straight line feature (represented by the coordinate of a cell in the Hough accumulator) the Bayesian approach is used. The HT is implemented by creating the accumulator array $H_C(\rho,\phi)$ (also called the Hough space) to represent each possible quantized set (ρ,ϕ). For each edge points (u,v) of the image, the line parameters $(\hat{\rho},\hat{\phi})$ are estimated and through (33) the probability that this point belongs to the line whose parameters are $\Theta_C = [\rho \quad \phi]^T$ is computed. The contribution of each edge point to all the possible image lines is obtained by iterating the computation of the edge point probability (33) for each possible set of (ρ,ϕ) in the discrete array. Each Θ_C of the accumulator denotes the hypothesis "*there exists a line whose parameters are (ρ, ϕ)*" and $\hat{\Theta}_{Ci}$, $i = 1, \cdots, n$ are conditional independent pieces of evidence concerning Θ_C and n is the total number of edge points. Hence $\hat{\Theta}_{Ci}$ is the event "*the i-th edge point belong to the line with parameters Θ_C*", $i = 1, ..., n$. Therefore the "*a posteriori* probability" of the line Θ_C given the evidences $\hat{\Theta}_{Ci}$, $i = 1, ..., n$ is specified as follows:

$$P(\Theta_C|\hat{\Theta}_{C1},\hat{\Theta}_{C2},...,\hat{\Theta}_{Cn}) = \frac{\frac{P(\Theta_C)}{P(\neg\Theta_C)} \prod_{i=1}^{n} \frac{P(\hat{\Theta}_{Ci}|\Theta_C)}{P(\hat{\Theta}_{Ci}|\neg\Theta_C)}}{1 + \frac{P(\Theta_C)}{P(\neg\Theta_C)} \prod_{i=1}^{n} \frac{P(\hat{\Theta}_{Ci}|\Theta_C)}{P(\hat{\Theta}_{Ci}|\neg\Theta_C)}}, \tag{34}$$

where $P(\Theta_C)$ is the prior probability about Θ_C, $P(\neg\Theta_C) := 1-P(\Theta_C)$, $P(\hat{\Theta}_{Ci}|\Theta_C)$ is the probability given by (33) and $P(\hat{\Theta}_{Ci}|\neg\Theta_C)$ can be deduced by the Bayes theorem

$$P(\hat{\Theta}_{Ci}|\neg\Theta_C) = \frac{P(\neg\Theta_C|\hat{\Theta}_{Ci})P(\hat{\Theta}_{Ci})}{P(\neg\Theta_C)}, \tag{35}$$

where $P(\neg\Theta_C|\hat{\Theta}_{Ci}) = 1 - P(\Theta_C|\hat{\Theta}_{Ci})$, with $P(\Theta_C|\hat{\Theta}_{Ci})$ specified as follows:

$$P(\Theta_C|\hat{\Theta}_{Ci}) = \frac{P(\hat{\Theta}_{Ci}|\Theta_C)P(\Theta_C)}{P(\hat{\Theta}_{Ci})}. \tag{36}$$

Substituting (36) in (35), the following relation is obtained

$$P(\hat{\Theta}_{Ci}|\neg\Theta_C) = \frac{P(\hat{\Theta}_{Ci}) - P(\hat{\Theta}_{Ci}|\Theta_C)P(\Theta_C)}{1 - P(\Theta_C)}. \tag{37}$$

Equations (34) and (37) allow us to compute the line probability for each bin Θ_C in the Hough accumulator.

After the edge points processing, the array is searched for peak elements. The peaks are local maxima. They identify the parameters of the most likely lines and their values exactly give the probability of these lines.

Note that, for each edge point, the updating of the probability value stored in the cells of the accumulator is not accomplished for all the cells (as in the standard HT), but only for those cells having line parameters linearly dependent. The complete correlation between ρ and ϕ reduces the computational efforts. In fact, for each detected edge point, the standard HT computes (31) for each $\Theta_C = [\rho \quad \phi]^T$ in $H_C(\rho, \phi)$.

4.2   Line Detection by Occupancy Grid

During the robot exploration, the value of the cells in the occupancy grid are updated using the probabilistic signal level fusion of sonars readings proposed in [2]. Straight line segments can be found in the occupancy grid as aligned cells with high probability of occupation. By interpreting a grid and its probabilities as an image with different level of intensity (grey level), it is possible to apply the HT to detect straight lines and to associate a probability at each detected line.

This probability is computed according to Bayesian and Soft Evidence theories [6]. Given a line with parameters vector $\Theta_G = (\rho, \phi)^T$, denote by n the number of cells in the occupancy grid belonging to the line Θ_G, by $c_i(\Theta_G)$, the event *"the i-th cell of the occupancy grid belonging to the line is occupied"* and by $\hat{c}_i(\Theta_G)$ the unsure event *"the i-th cell of the occupancy grid belonging to the line with vector parameters Θ_G is occupied with a proper uncertainty $P(c_i(\Theta_G)|\hat{c}_i(\Theta_G))$"*, $i = 1, \cdots, n$.

$P(c_i(\Theta_G)|\hat{c}_i(\Theta_G))$ is the probability of event $c_i(\Theta_G)$ given the evidence $\hat{c}_i(\Theta_G)$; it is the probability estimated by the sonars readings at the i-th cell of the occupancy grid and stored in the occupancy grid.

In the following, c_i and $\hat{c}_i$ are written without Θ_G argument for simplicity of notation.

Let $P(c_i) = P(\neg c_i) = 0.5$ be the prior occupancy/non occupancy probability of the i-th cell and denote with $P(\Theta_G)$ the prior probability of a line feature having parameters vector Θ_G. The Hough space $H_G(\rho, \theta)$ is used to store the existence evidence of each detected line feature. This way allows to find all the lines from the grid map, and to store the i-th cell of the grid map belonging to the line, the map coordinates and the probability $P(c_i|\hat{c}_i)$.

The existence evidence of a line feature depends on the evidence of all the cells (u_i,v_i) satisfying Eq. (9), with $u = u_i$ and $v = v_i$. Therefore the probability of each line feature with parameter Θ_G is the probability of the line conditioned to the events $\hat{c}_1, \hat{c}_2, \dots ,\hat{c}_n$; according to the Bayes theorem this probability has the form:

$$P(\Theta_G|\hat{c}_1,\hat{c}_2,...,\hat{c}_n) = \frac{\frac{P(\Theta_G)}{P(\neg\Theta_G)} \prod\limits_{i=1}^{n} \frac{P(\hat{c}_i|\Theta_G)}{P(\hat{c}_i|\neg\Theta_G)}}{1 + \frac{P(\Theta_G)}{P(\neg\Theta_G)} \prod\limits_{i=1}^{n} \frac{P(\hat{c}_i|\Theta_G)}{P(\hat{c}_i|\neg\Theta_G)}}, \tag{38}$$

where the terms $P(\hat{c}_i|\Theta_G)$, $P(\hat{c}_i|\neg\Theta_G)$ are stated in [8].

Each cell of the Hough space $H_G(\rho,\phi)$ is updated with the line probability computed using equation (38). Therefore a line segment on the grid map, with parameters Θ_G, is a local maximum in the Hough space with the probability $P(\Theta_G|\hat{c}_1,\hat{c}_2,\cdots,\hat{c}_n)$ greater than a probability threshold; in general the threshold is 0.5.

4.3 Fusion of Occupancy Grid Line Features and Digital Images Line Features

The proposed multisensor fusion process is defined as follows. All the lines detected from the portion of the occupancy grid corresponding to the portion of floor falling in the "visible space" are matched with the lines detected from the video image by projecting the lines of the video image on the floor plane.

The existence probability of the lines is stored in both Hough accumulators: $H_G(\rho,\phi)$ for the lines detected from the occupancy grid and $H_C(\rho,\phi)$ for the lines of the video image projected on the floor.

The matching algorithm is based on the combination of the lines probability encoded in both the Hough accumulators. A Bayesian estimator is developed. Each j-th line feature, described by its vector of parameters $\Theta_j=(\rho_j, \phi_j)$, has two probabilistic estimates $P_G=P(\Theta_{Gj}|\Theta_j)$, stored in H_G, and $P_C = P(\Theta_{Cj}|\Theta_j)$, stored in H_C, where Θ_{Gj} is the estimation of the line parameters Θ_j obtained by using the occupancy grid and Θ_{Cj} is the estimation of the line parameters Θ_j obtained by using the video data. Using the Bayes theorem, the combined estimate $P(\Theta_j|\Theta_{Gj} \cup \Theta_{Cj})$ is given by

$$P(\Theta_j|\Theta_{Gj} \cup \Theta_{Cj}) = \frac{P(\Theta_{Cj}|\Theta_j)P(\Theta_j|\Theta_{Gj})}{\sum\limits_{\Theta_j} P(\Theta_{Cj}|\Theta_j)P(\Theta_j|\Theta_{Gj})}. \tag{39}$$

By the Bayes theorem, $P(\Theta_j|\Theta_{Gj})$ is given as follows:

$$P(\Theta_j|\Theta_{Gj}) = \frac{P(\Theta_{Gj}|\Theta_j)P(\Theta_j)}{P(\Theta_{Gj})}. \tag{40}$$

By substituting (40) in (39), the following combination formula for fusing the sonar and video data is obtained:

$$P(\Theta_j | \Theta_{Gj} \cup \Theta_{Cj}) = \frac{\frac{P_C P_G}{P(\Theta_j)}}{\frac{P_C P_G}{P(\Theta_j)} + \frac{(1-P_C)(1-P_G)}{1-P(\Theta_j)}}, \tag{41}$$

that is also known as the Independent Opinion Pool [6].

4.4 Experimental Results

The proposed approach has been tested in an indoor environment by using the LabMate mobile base shown in Fig. 7. In this set of experiments the robot has been equipped with a proximity system composed of a half ring of 13 Polaroid ultrasonic sensors and with a low cost CCD web-camera Philips PCVC 675K. In the preliminary experimentation, the robot pose estimation has been performed by a simple odometric system. The camera for map building was installed in front of the vehicle and pointed down in the left side. Different experiments have been carried out, a sample of them is reported beneath.

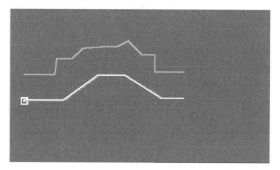

Fig. 10. Initial configuration (each cell with probability of 0.5) of the grid map with real obstacles (grey line), vehicle's trajectory (white line) and starting position (white box).

Figure 10 shows the indoor environment, the robot and the vehicle's starting position. Figure 11 shows the robot position and the occupancy grid map of the indoor environment during the robot movement, at a point belonging to the robot trajectory. The gray rectangle indicates the camera visible space. The map uses a grey scale, which goes from black (the null occupancy probability) to white (the maximum occupancy probability). In the robot position shown in Fig. 11, the video system acquires the image reported in Fig. 12, where the extracted line features are also displayed.

The related lines probability, stored in the Hough space H_C, are shown in Fig. 13. In this configuration, the part of occupancy grid map considered for line fusion is

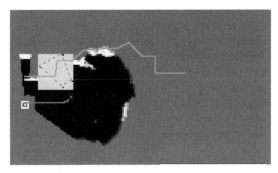

Fig. 11. Occupancy grid of the indoor environment built during the robot motion.

Fig. 12. Acquired digital image and extracted line features.

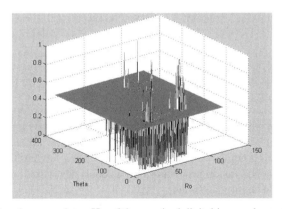

Fig. 13. Hough accumulator H_C of the acquired digital image shown in Fig. 12.

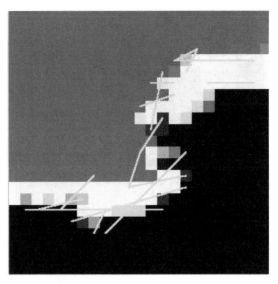

Fig. 14. Occupancy grid falling into the visible space of the camera and extracted line features.

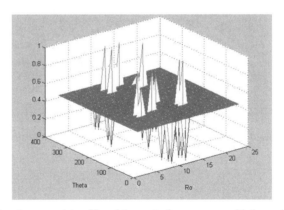

Fig. 15. Hough accumulator H_G of the part of occupancy grid shown in Fig. 14.

shown in Fig. 14, where the extracted line features are also reported. The related lines probability, stored in the Hough space H_G, are displayed in Fig. 15.

As shown in Figs. 12 and 14, and verified in a large set of experiments, the HT produces a high number of overlapping lines with a low probability values (see Figs. 13 and 15). The fusion procedure of the sonar data with the video data is able to extract only the significant lines.

The results of the fusion procedure (see Section 4.3) are reported in Fig. 16, where the lines specify the shape of the obstacles (walls) having over threshold probability. In this figure the dashed line represents the camera visible space. Finally,

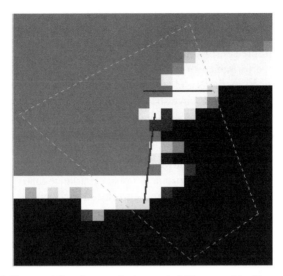

Fig. 16. Extracted line features fusing probability stored in H_C and H_G.

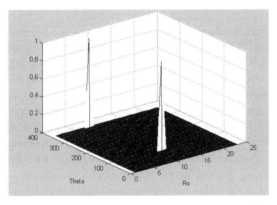

Fig. 17. Line features extracted fusing probability stored in H_C and H_G.

Figure 17 shows the probability associated to the extracted lines. The experiment shows the effectiveness of the proposed fusion technique for the straight line feature extraction from a real environment. Notice that preliminary set of experiments the robot pose estimation has been performed with a simple odometric system. This unavoidably affects the accuracy of the subsequent procedure for the extraction of the environment features. Future investigations will regard the integration of the proposed map building procedure with pose estimation algorithms based on a more accurate and complete sensor equipment. The expectation is a significantly improved accuracy of the localization process.

5 Conclusion

The use of a multiple sensor information introduces a significant improvement on the localization performance of a mobile robot thus enhancing its autonomy. In this chapter, different methods and techniques aimed at this purpose have been presented. The proposed methods have shown their ability in enhancing the localization capability of the robot and its capability of building up a reliable environment map. Different sensor equipments have been considered and a wide experimental validation has been performed. The proposed localization algorithms are based on the use of a linearized Kalman filter endowed with an adaptive algorithm for the on-line adjustment of the input and measurement noise covariance matrices.

The adaptation mechanism has been introduced to allow the filter to cope with realistic operating conditions. If the planned trajectory is relatively simple and not too long, some *a priori* engineered noise statistics may produce satisfactory results, but filter divergence may occur over complex trajectories. In this latter case the introduction of an adaptive algorithm seems to be the most effective and simple remedy. The experiments reported in this chapter confirmed that high performance of the localization algorithm is obtained in a wide range of real experimental situations.

The localization of a mobile robot requires some environment information, when this knowledge is *a priori* missing, it must necessarily be deduced from the sensor data. To this purpose, an algorithm has been proposed for the feature-based modeling of the environment. The algorithm is based on the fusion of sonar and video data. A probabilistic approach to the Hough Transform for extracting line features has been developed, and a Bayesian estimator has been introduced for matching the line features extracted by the video image with those extracted by the sonar image. The proposed technique has been shown to be able to build up large environment maps using line features. The resulting probabilistic model of the environment is simple and accurate, with a reduced memory demand. Experimental validation of the algorithm has been performed and satisfactory results have been obtained.

A very interesting and still open research field is the SLAM problem. It consists in defining a map of the unknown environment and simultaneously using this map to estimate the absolute location of the vehicle. An efficient solution of this problem appears to be of paramount importance because it would definitely confer autonomy to the vehicle. The most natural setting where this topic can be framed is the stochastic context of the Kalman filtering theory. The appealing features of this approach are:

i) the possibility of collecting all the available information and uncertainties of a different kind into a meaningful state-space representation,

ii) the recursive structure of the solution.

In this context, a natural way of dealing with the SLAM problem appears to be the definition of a stochastic state-space model whose state vector contains both the states of the vehicle model and the states of landmarks and map geometric features.

The work described in this chapter represents a solid basis of theoretical background and practical experience from which the numerous questions raised by this stimulating problem can be addressed.

References

1. B.D.O. Anderson and J.B. Moore, *Optimal Filtering*, Prentice-Hall, 1979.
2. A. Angeloni, T. Leo, S. Longhi, and R. Zulli, "Real time collision avoidance for mobile robots," *6th IMC Int. Symp. on Measurement and Control in Robotics*, pp. 239–244, 1996.
3. G.C. Anousaki and K.J. Kyriakopoulos, "Simultaneous localization and map building for mobile robot navigation," *IEEE Robotics and Automation Mag.*, vol. 6, no. 3, pp. 42–53, 1999.
4. N. Ayache and O.D. Faugeras, "Maintaining representations of the environment of a mobile robot," *IEEE Trans. on Robotics and Automation*, vol. 5, pp. 804–819, 1989.
5. M. Beckerman and E.M. Oblow, "Treatment of systematic errors in the processing of wide-angle sonar sensor data for robotic navigation," *IEEE Trans. on Robotics and Automation*, vol. 6, pp. 137–145, 1990.
6. E.A. Bender, *Mathematical Methods in Artificial Intelligence*, IEEE Computer Society Press, 1996.
7. A. Bonci, T. Leo, and S. Longhi. "Data fusion for visual and ultrasonic map-building," *Proc. of 6th IFAC Symp. on Robot Control*, pp. 241–256, 2000.
8. A. Bonci, T. Leo, and S. Longhi. "Ultrasonic and video data fusion for mobile robot navigation ," *Proc. of 10th Mediterranean Conf. on Control and Automation*, 2002.
9. M. Bonifazi, F. Favi, T. Leo, S. Longhi, and R. Zulli, "A developing environment for the solution of the navigation problem of mobile robots with non-holonomic constraints," *Proc. of 4th IEEE Mediterranean Symp. on New Direction in Control Automation*, pp. 107–112, 1996.
10. J. Borenstein and L. Feng, "Measurement and correction of systematic odometry errors in mobile robots," *IEEE Trans. on Robotics and Automation*, vol. 12, pp. 869–880, 1996.
11. J. Borenstein and Y. Koren, "Error eliminating rapid ultrasonic firing for mobile robot obstacle avoidance," *IEEE Trans. on Robotics and Automation*, vol. 11, pp. 132–138, 1995.
12. G. Bourhis, O. Horn, O. Habert, and A. Pruski, "An autonomous vehicle for people with motor disabilities," *IEEE Robotics and Automation Mag.*, vol. 7, no. 1, pp. 20–28, 2001.
13. Ö. Bozma and R. Kuc, "Characterizing pulses reflected from rough surfaces using ultrasound," *J. Acoustical Society of America*, vol. 89, pp. 2519–2531, 1991.
14. R.A. Brooks, "A robust, layered control system for a mobile robot," *IEEE Trans. on Robotics and Automation*, vol. 2, pp. 14–23, 1986.
15. R.G. Brown and P.Y.C. Hwang, *Introduction to Random Signals and Applied Kalman Filtering*, Wiley, 1997.
16. J.A. Castellanos, J. Neira, and J.D. Tardós, "Multisensor fusion for simultaneous localization and map building," *IEEE Trans. on Robotics and Automation,*, vol. 17, pp. 908–914, 2001.
17. L. Charbonnier and A. Fournier, "Heading guidance and obstacles localization for indoor mobile robot," *Proc. of 7th Int. Conf. on Advanced Robotics*, pp. 507–513, 1995.
18. H. Choset and K. Nagatani, "Topological simultaneous localization and mapping (SLAM): Toward exact localization without explicit localization," *IEEE Trans. on Robotics and Automation*, vol. 17, pp. 125–137, 2001.
19. G. Conte, S. Longhi, and R. Zulli, "Motion planning for unicycle and car-like robots," *Int. J. of Systems Science*, vol. 27, pp. 791–798, 1996.
20. J.L. Crowley, "World modeling and position estimation for a mobile robot using ultrasonic ranging," *Proc. of 1989 IEEE Int. Conf. on Robotics and Automation*, pp. 674–680, 1989.

21. A. Curran and K.J. Kyriakopoulos, "Sensor-based self-localization for wheeled mobile robots," *Proc. of 1993 IEEE Int. Conf. on Robotics and Automation*, vol. 1, pp. 8–13, 1993.
22. A.J. Davison and D.W. Murray, "Simultaneous localization and map-building using active vision," *IEEE Trans. on Pattern Analysis and Machine Intelligence*, vol. 24, pp. 865–880, 2002.
23. M. Di Marco, A. Garulli, S. Lacroix, and A. Vicino, "A set theoretic approach to the simultaneous localization and map building problem," *Proc. of 39th IEEE Conf. on Decision and Control,*, pp. 833–838, 2000.
24. M.W.M.G. Dissanayake, P. Newman, S. Clark, H.F. Durrant-Whyte, and M. Csorba, "A solution to the simultaneous lacalization and map building (SLAM) problem," *IEEE Trans. on Robotics and Automation*, vol. 17, pp. 229–241, 2001.
25. M.W.M.G. Dissanayake, P. Newman, H.F. Durrant-Whyte, S. Clark, and M. Csorba, "An experimental and theoretical investigation into simultaneous lacalozation and map building," in P. Corke and J. Trevelyan (Eds.), *Experimantal Robotics IV*, pp. 265–274, Springer Verlag, 2000.
26. M. Drumheller, "Mobile robot localization using sonar," *IEEE Trans. on Pattern Analysis and Machine Intelligence*, vol. 9, pp. 325–332, 1987.
27. R.O. Duda and P.E. Hart, "Use of the Hough transform to detect lines and curves in pictures," *Comm. ACM* vol. 15, pp. 11–15, 1972.
28. A. Elfes, "Sonar-based real-world mapping and navigation," *IEEE J. of Robotics and Automation*, vol. 3, pp. 249–265, 1987.
29. A. Elfes, "Sonar-based real world mapping and navigation," in J.I. Cox and G.T. Wilfgong (Eds.), *Autonomous Robot Vehicles*, Springer Verlag, 1990.
30. H.J.S. Feder, J.J. Leonard, and C.M. Smith, "Adaptive mobile robot navigation and mapping," *Int. J. of Robotics Research*, vol. 18, pp. 650–668, 1999.
31. F. Figueroa and A. Mahajan, "A robust navigation system for autonomous vehicles using ultrasonics," *Control Engineering Practice*, vol. 2, pp. 49–59, 1994.
32. S. Fioretti, T. Leo, and S. Longhi, "A navigation system for increasing the autonomy and the security of powered wheelchairs," *IEEE Trans. on Rehabilitation Engineering*, vol. 8, pp. 490–498, 2000.
33. K.-S. Fu, R.C. Gonzales, and C.S.G. Lee *Robotics: Control, Sensing, Vision and Intelligence*, McGraw-Hill, 1989.
34. G. Garcia, P. Bonnifait, and J.-F. Le Corre, "A multisensor fusion localization algorithm with self-calibration of error-corrupted mobile robot parameters," *Proc. of 7th Int. Conf. on Advanced Robotics*, pp. 391–397, 1995.
35. J.E. Guivant, F.R. Masson, and E.M. Nebot, "Simultaneous localization and map building using natural features and absolute information," *Robotics and Autonomous Systems*, pp. 79–90, 2002.
36. J.E. Guivant and E.M. Nebot, "Optimization of the simultaneous localization and map-building algorithm for real-time implementation," *IEEE Trans. on Robotics and Automation,*, vol. 17, pp. 242–257, 2001.
37. R.M. Haralick and L.G. Shapiro, *Computer and Robot Vision, Vol. 1*, Addison Wesley, 1992.
38. J. Heikkila and O. Silven, "A four-step camera calibration procedure with implicit image correction," *Proc. of 1997 IEEE Conf. on Computer Vision and Pattern Recognition*, pp. 1106–1112, 1997.
39. A.A. Holenstein, M.A. Müller, and E. Badreddin, "Mobile robot localization in a structured environment cluttered with obstacles," *Proc. of 1992 IEEE Int. Conf. on Robotics and Automation*, pp. 2576–2581, 1992.

40. P.V.C. Hough, *Methods and Means for Recognising Complex Patterns*, U.S. Patent, vol. 3, 069, 654, 1962.
41. A.H. Jazwinsky, *Stochastic Processes and Filtering Theory*, Academic Press, 1970.
42. L. Jetto, S. Longhi, and G. Venturini, "Development and experimental validation of an adaptive Extended Kalman Filter for the localization of mobile robots," *IEEE Trans. on Automatic Control*, vol. 15, pp. 219–229, 1999.
43. L. Jetto, S. Longhi, and D. Vitali, "Localization of a wheeled mobile robot by sensor data fusion based on a fuzzy logic adapted Kalman filter," *Control Engineering Practice*, vol. 7, pp. 763–771, 1999.
44. Q. Ji and R.M. Haralick, "An improved Hough transform technique based on error propagation," *1998 IEEE Int. Conf. on Systems Man and Cybernetics*, pp. 4653–4658, 1998.
45. A.C. Kak, "Depth perception for robots," in S.Y. Nof (Ed.), *Handbook of Industrial Robotics*, pp. 272–319, Wiley, 1985.
46. K. Kobayashi, K.C. Cheok, and K. Watanabe, "Fuzzy logic rule-based Kalman filter for estimating true speed of a ground vehicle," *Intelligent Automation and Soft Computing*, vol. 1, pp. 179–190, 1995.
47. R. Kuc and V.B. Viard, "A physically based navigation strategy for sonar-guided vehicles," *Int. J. of Robotics Research*, vol. 10, no. 2, pp. 75-87, 1991.
48. B.J. Kuipers and Y.T. Byun, "A robot exploration and mapping strategy based on a semantic hierarchy of spatial representations," *Robotics and Autonomous Systemes*, vol. 8, pp. 47–63, 1991.
49. H.J. Larson and B.O. Shubert, *Probabilistic Models in Engineering Sciences, Vol. 1: Random Variables and Stochastic Processes*, Wiley, 1979.
50. R.K. Lenz and R.Y. Tsai, "Techniques for calibration of the scale factor and image center for high accuracy 3D machine vision metrology," *IEEE Trans. on Pattern Analysis and Machine Intelligent*, vol. 10, pp. 713–720, 1988.
51. J.J. Leonard and H.F. Durrant-Whyte, "Mobile robot localization by tracking geometric beacons," *IEEE Trans. on Robotics and Automation*, vol. 7, pp. 376–382, 1991.
52. J.J. Leonard and H.F. Durrant-Whyte, *Direct Sonar Sensing for Mobile Robot Navigation*, Kluver Academic Publishers, 1992.
53. J.J. Leonard, H.F. Durrant-Whyte, and I.J. Cox, "Dynamic map building for an autonomous mobile robot," *Int. J. of Robotics Research*, vol. 11, pp. 286–298, 1992.
54. J.J. Leonard and H.J.S. Feder, "A computationally efficient method for large-scale concurrent mapping and localization," *Proc. of 9th Int. Symp. of Robotics Research*, pp. 169–176, 1999.
55. T.S. Levitt and D.T. Lawton, "Qualitative navigation fot mobile robots," *Artificial Intelligence J.*, vol. 44, pp. 305–360, 1990.
56. F. O'Gorman and M.B. Clowes, "Finding picture edges through collinearity of feature points," *IEEE Trans. on Computers*, vol. 25, pp. 449–454, 1976.
57. L. Ojeda, H. Chung, and J. Borestein, "Precision calibration of fiber-optic gyroscopes for mobile robot navigation," *Proc. of 2000 IEEE Int. Conf. on Robotics and Automation*, pp. 2064–2069, 2000.
58. E. Prassler, J. Scholz, and P. Fiorini, "A robotic wheelchair for crowded public environments," *IEEE Robotics and Automation Mag.*, vol. 7, no. 1, pp. 38–45, 2001.
59. B. Schiele and J.L. Crowley, "A comparison of position estimation techniques using occupancy grids," *Robotics and Autonomous Systems*, vol. 12, pp. 153–171, 1994.
60. C.C. Slama, *Manual of Photogrammetry*, American Society of Photogrammetry, 1980.
61. I.E. Sobel, *Camera Models and Machine Perception*, Ph.D. Thesis, Electrical Engineering Department, Stanford University, 1970.

62. S. Thrun, D. Fox, and W. Burgard, "A probabilistic approach to concurrent mapping and localization for mobile robots," *Machine Learning and Autonomous Robots*, vol. 31, pp. 29–53, 1998.

63. R.C. Tsai, "A versatile camera calibration technique for high accuracy 3D machine vision metrology using off-the-shelf TV cameras and lenses," *IEEE J. of Robotics and Automation*, vol. 3, pp. 323–344, 1987.

64. J. Vaganay, M.J. Aldon, and A. Fournier, "Mobile robot attitude estimation by fusion of inertial data," *Proc. of 1993 IEEE Int. Conf. on Robotics and Automation*, vol. 1, pp. 277–282, 1993.

65. P. van Turennout, G. Honderd, and L.J. van Schelven, "Wall-following control of a mobile robot," *Proc. of 1992 IEEE Int. Conf. on Robotics and Automation*, pp. 280–285, 1992.

66. C.M. Wang, "Localization estimation and uncertainty analysis for mobile robots," *Proc. of 1988 IEEE Int. Conf. on Robotics and Automation*, pp. 1230–1235, 1988.

67. S.B. Williams, G. Dissanayake, and H.F. Durrant-Whyte, "Towards terrain-aided navigation for underwater robotics," *Advanced Robotics*, vol. 15, pp. 533–550, 2001.

68. S.B. Williams, G. Dissanayake, and H. Durrant-Whyte, "An efficient approach to the simultaneous localization and mapping problem," *Proc. of 2002 IEEE Int. Conf. on Robotics and Automation* , pp. 406–411, 2002.

69. B. Yamauchi, A. Schultz, and W. Adams, "Mobile robot exploration and map building with continuous localization," *Proc. of 1998 IEEE Int. Conf. on Robotics and Automation*, pp. 3715–3720, 1998.

70. E. Zalama, G. Candela, J. Gómez, and S. Thrun, "Concurrent mapping and localization for mobile robots with segmented local maps," *Proc. of 2002 IEEE/RSJ Conf. on Intelligent Robots and Systems*, pp. 546–551, 2002.

71. G. Zunino and H.I. Christensen, "Simultaneous localization and mapping in domestic environments," *Proc. of Int. Conf. on Multisensor Fusion and Integration for Intelligent Systems*, pp. 67–72, 2001.

On The Problem of Simultaneous Localization, Map Building, and Servoing of Autonomous Vehicles

Antonio Bicchi, Federico Lorussi, Pierpaolo Murrieri, and Vincenzo Scordio

Centro Interdipartimentale di Ricerca "Enrico Piaggio"
Università di Pisa
Via Diotisalvi 2, 56125 Pisa, Italy
<bicchi,p.murrieri>@ing.unipi.it, <lorussi,scordio>@piaggio.ccii.unipi.it
http://www.piaggio.ccii.unipi.it

Abstract. In this chapter, we consider three of the main problems that arise in the navigation of autonomous vehicles in partially or totally unknown environments, i.e. building a map of the environment, self-localizing, and servoing the robot so as to achieve given goals based on sensorial information. As compared to most part of the existing literature on SLAM, we privilege here a system-theoretic view of the problem, which allows the localization and mapping problems to be cast in a unified framework with the control problem. The chapter is an overview of existing results in this vein, and of some interesting directions for research in the field.

1 Introduction

Autonomous vehicles have a wide range of applications, both in indoor and outdoor environments, and represent one of the areas with largest potential for advanced robotics. A very important trend in research related to mobile robots is concerned with their sensorization, and in particular with the tradeoffs between effectiveness and cost of different possible sensorial equipments.

Three of the main technical difficulties in applying mobile robots to partially or totally unstructured environments are indeed sensor-related: the localization of the vehicle with respect to the environment, the construction of a map of the environment itself, and the control of the vehicle to desired postures relative to the environment. Naturally, the three problems are closely interconnected. While the acronym SLAM (Simultaneous Localization And Map building) has been gaining wide acceptance in the robotics literature [5,28,40] to indicate the composition of the first two aspects, the connection to control is less frequently addressed. Indeed, in the SLAM literature, vehicles are often commanded in open loop. On the other hand, in the rather extensive literature on control of autonomous robots, localization is often simply taken for granted. Such is the case e.g. in many papers dealing with set-point stabilization of wheeled vehicles, which assume full state information, viz. [6,7,13,44,33,8,12]. In practical applications of automated vehicle control, however, one is confronted with the problem of estimating the current position and orientation of the vehicle only through indirect, noisy measurements by available sensors. Although much work has been done on techniques for vehicle localization based on combinations of sensory

B. Siciliano et al. (Eds.): Advances in Control of Articulated and Mobile Robots, STAR 10, pp. 223–242, 2004.

information (odometry, laser range finders, cameras, etc.), very little is known about the real-time connection of a localization algorithm and a feedback control law.

In this chapter, we consider the problem of simultaneous localization, mapping, and servoing (SLAMS) from a unified system-theoretic viewpoint, and report on work towards integrating solutions allowing an autonomous vehicle to navigate in an unknown environment. The chapter is organized as follows: in Section 2 we formulate the problem under consideration, and in Section 3 we provide a brief survey of the state of the art. In Section 4 we discuss aspects related to the existence of solutions to the SLAM problem, and to the choice of optimal exploratory paths to elicit SLAM information. In Section 5 we report on the problem of simultaneous localization and servoing, before concluding in Section 6.

2   Modeling of the SLAMS Problem

Let us consider a system comprised of a vehicle moving in an environment with the aim of localizing itself and the environment features. For simplicity, we assume that features are distinctive 3D points in the environment where the vehicle moves (more general features are described e.g. in [38]). The vehicle is endowed with sensors, such as a radial laser rangefinder or video cameras. Both the vehicle initial position and orientation, and the feature positions, are unknown or, more generally, known up to some a priori probability distribution. A particular pose of the vehicle, or set of poses, is regarded as the goal. Sensor readings corresponding to the goal pose are known (by e.g. recording them in a preliminary learning phase). Among the features that the sensor head detects in the robot environment, we will distinguish between those belonging to objects with unknown positions (which we shall call *targets*), and those belonging to objects whose absolute position is known (which will be referred to as *markers*). Indeed, as it can be argued, this distinction is only useful for simplicity of description, as in general the case is that there exist features that are more or less uncertain.

The vehicle dynamics are supposed to be slow enough to be neglected (dynamics do not add much to the problem structure, while increasing formal complexity). Kinematics of wheeled vehicles can usually be written as a nonlinear system of the type $\dot{x} = G(x)u$, where $x \in \mathbb{R}^{n_v}$ is the robot pose (typically, $n_v = 3$ for a vehicle moving in a plane with an orientation), and $u \in \mathbb{R}^m$ are the input velocities. It is often the case where the system velocities are affected by disturbances μ (such as slippage of the wheels), and the model is accordingly modified to include process noise as $\dot{x} = G(x)(u + \mu)$.

Let the i–th target absolute coordinates be denoted by $p_i \in \mathbb{R}^d$, with $d = 2$ for planar features and $d = 3$ in case of 3D environments, and use $p \in \mathbb{R}^{dn_f}$ to denote the collection of all features. According to the sensor equipment specifics, the relative position of the vehicle and of the features form sensor readings, or *observables*, described by the map $h : \mathbb{R}^{n_v} \times \mathbb{R}^{dn_f} \to \mathbb{R}^q$, $(x, p) \mapsto y = h(x, p)$. Measurement noise ν adds to this as $y = h(x, p) + \nu$.

In system-theoretic terms, the three problems in SLAMS can be described by referring to the input-state-output system

$$\begin{bmatrix} \dot{x} \\ \dot{p} \end{bmatrix} = f(x, p, u, \mu) = \begin{bmatrix} G(x) \\ 0 \end{bmatrix} (u + \mu) \tag{1}$$
$$y = h(x, p) + \nu.$$

In this framework, localization and mapping are *observability* problems, dealing with the reconstruction of the present pose x and feature map p, respectively, from current and past observables, from model and input knowledge, and from statistics on process noise μ and measurement noise ν. Servoing is a *stabilization* problem, aiming at devising what inputs u are to be given to the system so as to reach the desired pose, based on available data. Should the current pose x be known exactly at all times, servoing would amount to find a state feedback law in the form $u(x, t)$, such that $\dot{x} = G(x)u(x, t)$ asymptotically converged to the desired pose. However, such knowledge is not available in general, because typically $q < n_v + dn_f$ and, even when this inequality would be reversed (such as when using absolute landmarks and a trinocular stereo camera head), because of measurement noise. Servoing in SLAMS should therefore be regarded in general as an *output stabilization* problem, whereby a new dynamic system must be designed in the additional states $w \in \mathbb{R}^r$ as

$$\dot{w} = S(w, y) \tag{2}$$
$$u = F(w, y)$$

such that, when connected to system (1), asymptotic stability of the compound $n_v + dn_f + r$ states can be achieved. It is often (but not always) the case that the auxiliary system (2) includes an *estimator* of the system (1), i.e. its design is aimed at achieving the convergence of $w(t)$ to the pose $x(t)$ (the prevailing design for the estimator is based on Extended Kalman Filters, see below). According to this approach, a design is often attempted for the control in the form of a state-feedback stabilizer $u(w, t)$, where w is used in place of x. Naturally, convergence of the estimator and of the state-feedback law separately are only necessary conditions in order for their composition to provide a stable and satisfactory behavior.

The model in (1) is sometimes referred to as *world-centric*. It is rather obvious that, unless geographic markers or other equivalent information (from compass, GPS, etc.) are present, reconstruction of absolute robot position and orientation is impossible. A different description of the same problem can hence be given in coordinates relative to the vehicle (a *robot-centric* model), which would be written in the form

$$^v\dot{p} = \ell(^vp, u, \mu) \tag{3}$$
$$y = \hat{h}(^vp) + \nu.$$

Such a model is applicable for instance to the case where a camera is mounted on the vehicle, with the output map $\hat{h}(\cdot)$ representing the projection of 3D features to the image plane of the camera. Output feedback control of (3) amounts then to what is commonly referred to as image-based visual servoing of the vehicle. In this case, explicit estimation of the robot pose is clearly unnecessary.

In the rest of this chapter, we will discuss these different aspects of the SLAMS problem in more detail, emphasizing the insight that an integrated system theoretic approach can bring to the field.

3   Approaches to the SLAM Problem

As shown in the former section, the SLAM problem —also known in the literature as CML (Concurrent Mapping and Localization)— is characterized by two sources of uncertainty: the vehicle model (because of both uncertain parameters appearing in the dynamics and process noise) and sensor noise.

Uncertainty can be dealt with in basically two ways, i.e. deterministically or by using probabilistic models. The first approach assumes that all uncertainty sources may generate errors that are unknown but bounded, and seeks for bounds on how these error can propagate through the reconstruction process. Naturally, the problem tends to be overly complex from the computational and memory-occupation viewpoints; hence efficient algorithms to approximate the worst-case bounds are in order. An application of this approach to robot localization is reported in [18], where an efficient, recursive algorithm to approximate the set of robot poses compatible with present and past measurements is presented.

Deterministic algorithms tend to suffer from excessive conservativeness, and are typically not very suited to take into account the existence of large, sporadic errors in sensor readings (*outliers*), which are common in some types of sensors used in SLAMS (e.g. spurious reflections of lasers or sonars, feature mismatch, etc.). When an excess of conservatism is not justified by particularly risk-sensitive applications, it is often preferred to adopt probabilistic models of uncertainty.

The basis for virtually all probabilistic methods is Bayesian theory of inference, which assumes that the statistical properties of the *data space* and of the *model space* are well defined. These are the vector spaces, of suitable dimension, where observables y and unknowns (and estimates thereof, denoted for brevity as x) take their values, and where a probability density function (p.d.f.) is defined for the variables of interest. The *a priori* state of information consists in a p.d.f. defined over the model space X, $f_{prior}(x)$, which models any knowledge one may have on the system model parameters independently from the present act of measurement, due e.g. to physical insight or to independent measurements carried out previously.

In the formation of estimates, two information sources are to be considered, i.e. the forward solution of the physical model, and the act of measuring itself. The state of information on the experimental uncertainties in measurement outputs can be modelled by means of a p.d.f. $f_{exp}(y)$ over the data space Y (this should be provided by the instrument supplier), while modelling errors (due to imperfection of (1), or to process noise) can be represented by a conditional p.d.f. $f_{mod}(y|x)$ in the data space Y (or, more generally, by a joint p.d.f $f_{mod}(y, x)$ over $X \times Y$).

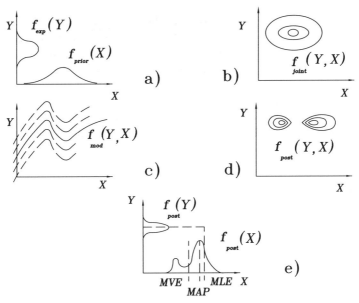

Fig. 1. The process of Bayesian inference. A priori information on the model space, $f_{prior}(x)$, and information on experimental data $f_{exp}(y)$ are independent (a), and combine in the joint p.d.f. $f_{joint}(y,x)$ (b). Information on modelling is represented by $f_{mod}(y,x)$ (c). The conjunction of $f_{joint}(y,x)$ and $f_{mod}(y,x)$ is $f_{post}(y,x)$ (d). The marginal p.d.f.'s $f_{post}(x)$ and $f_{post}(y)$ can be obtained directly from $f_{post}(y,x)$. Different estimators can be applied to these results, as illustrated in (e).

Fusing the different information in an estimate of x leads to a *posterior* p.d.f over X, that is described by Bayes formula

$$f_{post}(x) = f(x|y) = \alpha_b\, f_{prior}(x) \int_Y f_{exp}(y) f_{mod}(y|x) dy, \qquad (4)$$

where α_b is a normalization factor such that $\int_X f_{post}(x)dx = 1$. The process of information fusion is described in Fig. 1 (adapted from [43]), with reference to the case where the measurement equation forming observables y from unknowns x is nonlinear (such as it actually is in SLAM). Although the posterior p.d.f. on the model space represents the most complete description of the state of information on the quantity to be measured one may wish, a final decision on what is the "best" estimate of x needs usually be taken. Several possibilities arise in general, such as the *maximum a posteriori* estimate (MAP), *maximum likelyhood* estimates (MLE, which coincides with MAP if no priors are available), the *minimum variance* estimate (MVE) alias *minimum mean square* (MMSE). Figure 1e illustrates these estimates. While very little can be said in general about the performance of such estimators, well known particularizations apply under certain assumptions on the prior distributions. Thus, if a normal distribution (an order-2 Gaussian) can be assumed for all prior information, the MAP estimate enjoys many useful properties; among them, perhaps most importantly for the problem at hand), since the convolution in (4) of two Gaussian distributions is Gaussian, the modelling and experimental errors in measurements simply combine by addition of the covariance matrices of experimental and modelling errors, $C_Y = C_{exp} + C_{mod}$. Roughly speaking, errors in the model knowledge (kinematic model of the systems and odometry errors) can be ignored, provided that experimental measurement errors in y are suitably increased. This result holds for nonlinear sensor models as well. For linearized measurement models ($y = Hx$), the a posteriori p.d.f. would also be Gaussian, the MVE and MAP estimates would coincide and can be evaluated as

$$\hat{x} = C_{post}(H^T C_Y^{-1} y + C_{prior}^{-1} x_{prior}),$$
$$C_{post} = (\mathcal{F} + C_{prior}^{-1})^{-1}, \qquad (5)$$

where $\mathcal{F}$, the *Fisher information matrix* for the linear case at hand, is defined as

$$\mathcal{F} = H^T C_Y^{-1} H. \qquad (6)$$

As a final remark, the Gauss-Markov theorem [37] ensures that the estimate (5) is the Best Linear Unbiased Estimate (BLUE) in the minimum-variance sense even for non-Gaussian a priori distributions. This result may seem to indicate some "absolute optimality" of the least-squares estimate. However, the MVE of a non-Gaussian distribution may not be a significant estimate, as apparent in Fig. 1e. This is the case for instance when a few measurements are grossly in error (*outliers*): the MVE in this case can provide meaningless results. This fact is sometimes used to point out the *lack of robustness* of the MVE.

In the literature on mobile robot localization and mapping, methods to evaluate an estimate of the posterior p.d.f. over the space of unknown robot poses and targets have been studied extensively. While for an exhaustive review the reader is referred to [40], we limit ourselves to point out that methods proposed so far can be roughly classified in two main groups: batch and recursive.

Batch methods attempt as accurate a solution of the posterior as possible, by taking into account that often in SLAM the posterior p.d.f. is a complex multimodal distribution. To such complexity contribute different factors, among which the non-linearity of dynamics and measurement equations (1), and the fact that measurement noise in different measurements is statistically correlated, because errors in control accumulate over time, and they affect how subsequent measurements are interpreted ([40]). A crucial aspect of SLAM is indeed that, when features are not distinctive, multiple correspondences are possible, a problem also known as *data association*. The correspondence problem, consisting in determining if sensor measurements taken at different times correspond to the same physical object in the world, is very hard to be tackled, since the number of possible hypotheses can grow exponentially over time. A family of methods recently introduced to deal with these problems, which is based on Dempster's Expectation Maximization (EM) Algorithms [14,40], represent the current state of the art in this regard. However, since EM have to process data multiple times they are not suitable to real-time implementation, as needed e.g. to interface with servoing algorithms.

On the other hand, most often new updates of model estimates are needed in real time, without referring to the whole history of sensed data. To cope with this requirement, further simplifications are usually done: for instance, assuming a Gaussian posterior distribution, the given record of data can be completely described by the mean vector and the covariance matrix. When a new datum is available, all prior information can be extracted from those statistics. A method that does not use prior information explicitly, but through its statistics only, is called *recursive*. The Kalman filter is one such recursive method, implementing the optimal minimum variance observer for a linear system subject to uncorrelated, zero-mean, Gaussian white noise disturbances.

Unfortunately, these assumptions are not fulfilled in SLAM applications. Hence, different simplifying assumptions and approximations are employed. Filters resulting from repeated approximate linearization of (1) are commonly referred to Extended Kalman Filters (EKF). Although EKF's for the SLAM problem do not guarantee any optimality property, they remain the most widely used filters in SLAM. EKF maintain all information on the estimated posteriors in the vector of means and in a covariance matrix, whose update at each step is a costly operation (quadratic with the number of features). In practical implementations, a key limitation of EKF is the low number of features it can deal with.

Algorithms have been recently proposed to overcome this limitation. The Fast-Slam [32] algorithm is based on the assumption that the knowledge of the robot path renders measurements of individual markers independent, so that the problem of determining the position of K features could be decomposed into K estimation problems, one for each feature [32]. Compressed EKF (CEKF), see [20], stores and maintains all the information gathered in a local area with a cost proportional to the square of the number of landmarks in the area. This information can then be transferred to the rest of the global map with a cost that is similar to full SLAM, but in only one iteration. Sparse Extended Information Filter (SEIF), see [42], is an algorithm whose updates require constant time, independent of the number of features in the map. It exploits the particular form of the information matrix, i.e. the inverse of the covariance matrix. Since the information matrix is sparse, it possesses a large number of elements whose values, when normalized, are near zero and can be neglected in the updating process. Some algorithms, see [17,25], based on incremental update of uncertain maps, use a fuzzy logic approach to manage uncertainty on obstacle poses and successively implement obstacle avoidance strategies.

An interesting possibility in SLAM is the possibility of using multiple vehicles in a cooperative way in order to perform tasks more quickly and robustly than a single vehicle can do. In [15,41], the problem of performing concurrent mapping and localization with a team of cooperating autonomous vehicles is considered, and the advantages of such a multiagent cooperation are illustrated.

One of the most challenging topics in SLAM is the optimization of autonomous robotic exploration. Indeed, it is often the case that robots have degrees of freedom in the choice of the path to follow, which should be used to maximize the information that the system can gather on the environment. The problem is clearly of great relevance to many tasks, such as surveillance or exploration. However, it is in general a difficult problem, as several quantities have to be traded off, such as the expected gain in map information, the time and energy it takes to gain this information, the possible loss of pose information along the way, and so on. This problem is considered in detail in the next section.

4   Solvability and Optimization of SLAM

As already mentioned, simultaneous localization and mapping amounts to estimating the state of system (1) through integration of input velocities (odometry) and knowledge of the observations y. Input velocities and observables are affected by process and measurement noise, respectively.

We start by observing that system (1) is nonlinear in an intrinsic way, in the sense that approximating the system with a linear time-invariant model destroys the very property of observability: this entails that elementary theory and results on linear estimation do not hold in this case.

The intrinsic nonlinear nature of the problem can be illustrated directly by the simple example in Fig. 2 of a planar vehicle ($n_v = 3$) with M markers and N targets (hence $dn_f = 2N$). Outputs in this examples would be the $q = M + N$ angles

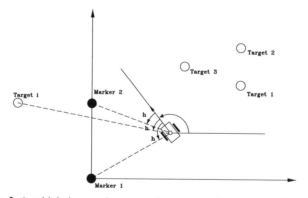

Fig. 2. A vehicle in an unknown environment with markers and targets.

formed by the rover fore axis with lines through the sensor head and the M markers and N targets. The linear approximation of system (1) at any equilibrium $x = x_0$, $p = p_0$, $u = 0$ would indeed have a null dynamic matrix

$$A = \left. \frac{\partial f(\cdot)}{\partial (x, p)} \right|_{eq.} = 0 \in \mathbb{R}^{(2N+3) \times (2N+3)}$$

and output matrix

$$C = \left. \frac{\partial h(\cdot)}{\partial (x, p)} \right|_{eq.} \in \mathbb{R}^{(M+N) \times (2N+3)}.$$

Hence, in any nontrivial case (i.e., whenever there is at least one target ($N \neq 0$) or there are less than three known markers ($M < 3$)) the linearized system is unobservable.

On the other hand, it is intuitively clear (and everyday's experience in surveyors' work) that simple triangulation calculations using two or more measurements from different positions would allow the reconstruction of all the problem unknowns, except at most for singular configurations. Analytically, complete observability of system (1) can be checked, as an exercise in nonlinear system theory, by computing the dimension of $< f(\cdot) \,|\, \mathrm{span}\,\{dh(\cdot)\} >$, the smallest codistribution that contains the output one-forms and is invariant under the control vector fields (see [2] for details on calculations). By such nonlinear analysis, it is also possible to notice that observability can be destroyed by choosing particular input functions, the so-called "bad inputs". A bad input for our example is the trivial input $u = 0$: the vehicle cannot localize itself nor the targets without moving. Other bad inputs are illustrated in Fig. 3.

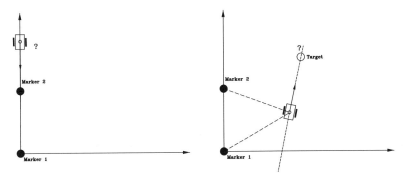

Fig. 3. A vehicle triangulating with two markers cannot localize itself if the inputs are such that it remains aligned with the markers; it cannot localize a target if it aims at the target directly.

In order to drive a rover to explore its environment, it is clear that bad inputs should be avoided. Indeed, the very fact that there exist bad inputs suggests that there should also be "good", and possibly optimal, inputs. To find such optimal exploratory startegies, however, the differential geometric analysis tools such as those introduced above are not well suited, as they only provide topological criteria for observability. What is needed instead is a metric information on the "distance" of a system from unobservability, and to how to maximize it. More generally, it is to be expected that different trajectories will elicit different amounts of information: a complete SLAM system should not only provide estimates of the vehicle and feature positions, but also as precise as possible a description of the statistics of those estimates as random variables, so as to allow evaluation of confidence intervals on possible decisions.

To provide a better understanding of how two different states can be distinguished via dynamic measurements, let us consider the output $y(t) = h(x, p) = y(x_o, u, t)$ as a function of the initial conditions x_o and of the inputs u. Let x_o^o and x_o' denote two different initial conditions, with $\|x_o^o - x_o'\| < \epsilon$, and let us consider

$$y(x_o', u, t) - y(x_o^o, u, t) = \left.\frac{\partial y}{\partial x_o}\right|_{x_o = x_o^o} (x_o' - x_o^o) + O^2(\epsilon) \tag{7}$$

i.e. a linear measurement equation of the form

$$\tilde{y}(t) + \delta_y = M(t)\tilde{x} \tag{8}$$

where $\tilde{x} = (x_o' - x_o^o)$ is unknown, $\tilde{y}$ comes from measurements, and the perturbation term δ_y accounts for measurement noise and approximation errors. Notice explicitly that the linear operator $M = \left.\frac{\partial y}{\partial x_o}\right|_{x_o = x_o^o}$ $\mathcal{F}$ depends in general on applied inputs, as only for very special systems (in particular, linear) superposition of effects of initial states and inputs holds. By premultiplying both sides of (8) by $M^T W$, with $W > 0$ a suitable positive definite matrix weighing accuracy of different sensors, and by integrating from time 0 to T, we obtain

$$\mathcal{Y} + \Delta_y = \mathcal{F}\tilde{x}, \tag{9}$$

where $\mathcal{Y} = \int_0^T M^T(t)W\tilde{y}(t)dt$, and $\mathcal{F} = \int_0^T M^T(t)WM(t)dt$ is the *Fisher information matrix* for our system.

Singularity of $\mathcal{F}$ (for some input choice) clearly implies that distinct initial values of the state exist which provide exactly the same measurements over the time interval, hence is tantamount to unobservability of the system.

A different argument to support the same conclusion can be derived from Kalman estimation theory. Indeed, in the linear case, for the covariance matrix P of a Kalman filter, the Cramèr-Rao inequalities [37] hold:

$$[\mathcal{F} + \mathcal{N}^{-1}]^{-1} \le P \le \mathcal{F}^{-1} + \mathcal{N} \tag{10}$$

where $\mathcal{F}$ is the Fisher information matrix (defined in (9) in this framework), and $\mathcal{N}$, the covariance matrix of process noise, is assumed to be independent of the trajectory. According to this, minimization of $\mathcal{F}^{-1}$ can be considered as a means to minimize P. This is further justified by the fact that, in the absence of process noise and of prior information, the Riccati equation solution for the filter is exactly $P(t) = \mathcal{F}^{-1}(t)$. Cramèr-Rao bounds can be extrapolated to estimate covariance for nonlinear systems (see e.g. [26]) (although, in the context of nonlinear systems, minimum-variance estimates do not enjoy the properties that make them desirable for linear systems, and MVE-based optimal sensor design is often questionable [1]).

From the above considerations on state reconstruction and on Cramèr-Rao inequalities, it is clear that the information matrix can provide the desired notion of "distance" from unobservability, that is, a merit figure for different inputs (hence trajectories) of the exploring rover. Indeed, the smallest eigenvalue $E = \lambda_{min}(\mathcal{F}) = 1/\|\mathcal{F}^{-1}\|_2$, the determinant index $D = {}^{(n_v + dn_f)}\sqrt{\det \mathcal{F}}$ the trace index $T = \frac{\text{trace } (\mathcal{F})}{n_v + dn_f}$, and the average-variance index $A = \frac{n_v + dn_f}{\text{trace } (\mathcal{F}^{-1})}$ are among the most often used such criteria (known as E– , D– , T–, and A–criterion, respectively).

Notice that information-based criteria do not reflect any particular choice in the estimator or filter adopted in the actual localization procedure, rather it is intrinsic to the reconstructibility of the state from the given trajectory. This is a very useful property, in view of the fact that several different estimators and filters can be applied to the SLAM problem.

The problem of choosing exploratory paths of fixed length L to maximize SLAM information can be formalized (in the E–criterion sense) as an *optimal control* problem, i.e.

$$\text{maximize } J(u) = \lambda_{min} (\mathcal{F}), \qquad (11)$$

subject to the constraints

$$L = \int_0^T \sqrt{(\dot{x}_1^2 + \dot{x}_2^2)} \, dt$$

$$\dot{x} = G(x)u; \quad x(0) = x_o$$

$$y = h(x).$$

Solving this problem can be expected to be quite difficult in general. Using system–theoretic tools, an analytic solution was given in [30] for the simplified case of an omnidirectional vehicle moving in a planar environment with only two markers. Extremal paths for the functional J were shown to be contained in the pencil of curves spanned by the parameter α as

$$\left[\cos(\alpha) \; \sin(\alpha) \right] \left(\frac{\partial y}{\partial x_o}^T \frac{\partial y}{\partial x_o} - \frac{\partial y}{\partial x_o}\bigg|_{x=x_o}^T \frac{\partial y}{\partial x_o}\bigg|_{x=x_o} \right) \begin{bmatrix} \cos(\alpha) \\ \sin(\alpha) \end{bmatrix} = 0, \quad (12)$$

where the actual value of α depends on L. It can be easily seen that the obtained pencil is a set of conics (some examples of optimal exploratory paths, for different lengths, are represented in Fig. 4).

Extensions of the analytic solutions to nonholonomically constrained vehicles with unknown target features are feasible (work in this direction is undergoing). However, to obtain solutions in most general cases, efficient numerical methods are in order. In a recent overview [40], where the importance of the SLAM optimization problem is acknowledged, currently available solutions are reported to be mostly limited to heuristic, greedy algorithms. Furthermore, most known methods often disregard the nonlinear character of the SLAM problem, which on the contrary is of large momentum, as we discussed.

The main limitation of gradient-descent methods in this framework is of course the presence of local minima in the information return function: application of methods from receding-horizon optimal control theory in this context can be expected to

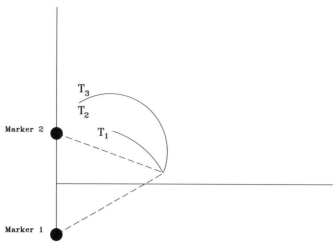

Fig. 4. Optimal trajectories for three different path lengths $T_1 = 1$ s, $T_2 = 2$ s, and $T_3 = 3$ s

offer a substantial edge. In the following, we illustrate application of such techniques to a few examples of on-line, numeric SLAM trajectory optimization.

To apply numerical methods, continuous-time system equations (1) or (3) are first discretized, so that the information matrix is rewritten as the sum of products

$$\mathcal{F} = \sum_{i=0}^{k} \frac{\partial y}{\partial x_o}\bigg|_{x_i}^{T} \frac{\partial y}{\partial x_o}\bigg|_{x_i} \tag{13}$$

evaluated at each point of a candidate trajectory. Using techniques developed in [35], we furthermore introduce a quantization of the input space (i.e., the set of possible incremental moves of the vehicle), thus inducing a discretization of the configuration space. It can be shown that, for vehicles with chained-form kinematics, the reachable set is indeed a lattice in this case, which is a very convenient structure to apply numerical search methods to.

If d is the cardinality of the input set, there are d^k paths of length k stemming from a generic configuration, for which the contribution to information is given by (13). An exhaustive search of the most informative path is possible for moderate values of d and k. The receding-horizon optimal control policy consists then in applying only the first control of the locally optimizing sequence, to recompute the next optimizing sequence, thus proceeding iteratively. The method can be easily used in conjunction with other techniques for e.g. obstacle avoidance. How practical the method depends very much on the affordable horizon length for which real-time computations are feasible; hence choices concerning time and input quantization, information representation, etc., are an important area of research.

Simulation results reported in Fig. 5 compare the performance of a greedy algorithm with the receding–horizon method. Walls are considered here as pure obstacles, i.e. they are detected if and only if the vehicle "bumps" into them, while information for self localization and mapping is only extracted from measurements relative to two markers (black circles) and to four target features. Results show how the receding-horizon methods collects richer information in this case.

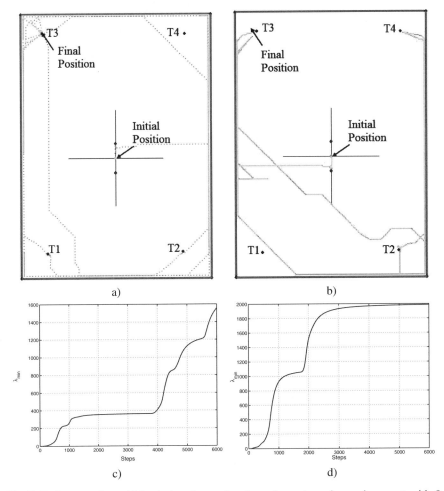

Fig. 5. Trajectory of a vehicle during the exploration of a rectangular environment with 2 markers and 4 target features, using gradient-descent (a) and a 3-steps receding horizon (b), respectively. Time evolutions of the corresponding information return function $E = \lambda_{min}(\mathcal{F})$ are reported in c) and d).

More simulation results relative to different environments are reported in Fig. 6. While these results show how the method is quite versatile in navigating in a cluttered environment fetching for information where that is available, it is of course an open research issue to provide a provable, quantitative assessment of the advantages of

this method with respect to others, and to design the numerous parameters that play
an important role in its implementation.

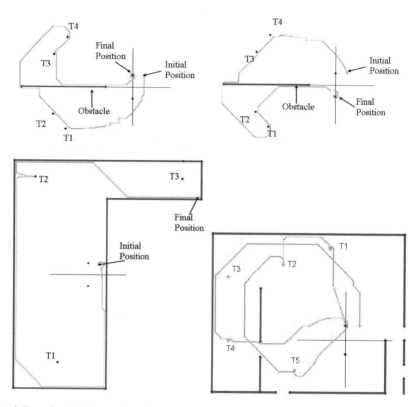

Fig. 6. Receding-horizon optimal trajectories in different environments, whereby the task of
maximizing the information return function leads the vehicle to cover target areas. Observe
how slightly different initial conditions may lead to completely different exploration strategies
(upper right and left), however with similar characteristics. More complex environments are
also dealt with satisfactorily (bottom left and right).

5   Simultaneous Localization and Servoing

As mentioned in the introduction, one of the consequences of the intrinsic nonlinearity of the SLAMS problem is that there exists no guarantee that, even assuming
that a converging estimator *and* a stabilizing state-feedback law are available, their
connection will provide an overall controller which behaves as expected.

Many different approaches can be taken to address this problem, depending on
the specifics of the task and of the models at hand. In this section, we will survey
two methods which differ in the generality they afford.

5.1 Observer-Based SLAMS

A first approach to design simultaneous localization and servoing for system (1) relies on techniques for output-feedback stabilization of nonlinear systems, based on nonlinear observers and extensions of the separation principle. An observer providing an estimate $\hat{x}$ of the current state x is said to enjoy the separation property if, whenever there exists a nominally stabilizing static state feedback $u(x)$, application of the control law $u(\hat{x})$ permits to achieve (local) stability of the closed-loop system.

The widely used EKF (see e.g. [5,19,31]) lacks in general provable properties of convergence and separation. Several authors underscored that application of EKF to localization data is often troublesome. Indeed, the filter convergence properties are very much prone to initialization of filter parameters (e.g., measurement and process covariances). Several alternative schemes of nonlinear observers have been proposed in the literature.

One way of designing an observer is to transform the original nonlinear system into another one for which the design is known. Transformations, which have been proposed in the literature, include system immersion [16], which permits obtaining a bilinear system if the observation space is finite dimensional, and linearization by means of output injection [21,27,29], which assumes that particular differential-geometric conditions on the system vector fields are verified. A nonlinear observer and its practical implementation have been presented in [9,4], in which the first step is writing the input affine nonlinear system in a so-called normal observation form. However, this form requires that the trivial input is an universal input [3] for the system. In our problem, this condition is violated by the mobile robot kinematics.

An interesting possibility for an estimator for the localization problem is the extension of the Luenberger filter in a nonlinear setting, by using the time derivatives of the input [45,39]. In [10], a local nonlinear observer for mobile robot localization was designed, based on the concept of Extended Output Jacobian (EOJ) matrix, which is the collection of the covectors associated to the considered elements of the observability space, i.e. the output and its derivative. Output derivatives are estimated by using high-pass filters. Local practical stability of the observation error dynamics is guaranteed since persistent perturbations introduced by filters can be made arbitrarily small. A singularity-avoidance exploration task is also addressed to deal with the singularity occurrences in the EOJ matrix. For such an observer scheme, [10] showed a local separation property to hold.

Observer-based approaches are applicable to rather general models of vehicles and sensorial equipment, but has correspondingly some weaknesses. In the first place, convergence and separation can only be proven locally. While this is to be expected with complex nonlinear systems such as those at hand, the drawback shown by laboratory practice is that the large initial estimation errors or displacements from the desired pose often prevent correct functioning of the system.

5.2 Visual Servoing

When the SLAMS problem is specialized for a particular class of sensors and vehicles, more powerful techniques can usually be devised. In this paragraph, we report

Fig. 7. Image grabbed from the robot camera at the target position, with four selected control features.

on a particular but important case, where we assume that the sensorial information consists of a video camera mounted on-board the vehicle.

Visual servoing techniques, which have been profitably used in recent years mostly for the control of robot arms [23], use visual information either directly, by the computation of an image error signal, or indirectly, by the evaluation of the state of the system. These two approaches were classified by Weiss in 1984 as *Image Based Visual Servoing* (IBVS) and *Position Based Visual Servoing* (PBVS), respectively. Indeed, these two schemes should be regarded as the end points of a range of different possibilities, whereby the raw sensorial information is gradually abstracted away to a more structured representation using some knowledge of the robot-environment model (a scheme which is roughly half-way between IBVS and PBVS was used e.g. in [11]).

IBVS and other sensor-level control schemes have several advantages, such as robustness (or even insensitivity) to modelling errors and hence suitability to unstructured scenes and environments. On the other hand, PBVS and in general higher-level control schemes also have important attractive features. Using the PBVS approach, for instance, the control law can be synthesized in the usual working coordinates for the robot, and thus usually a simpler synthesis is made possible. Furthermore, abstracting sensor information to a higher level of representation allows using different sensorial sources. In the example of a camera mounted on a mobile robot, for instance, the synergistic use of odometry and visual feedback is only possible if this information can be taken to some common denominator where they can be fused coherently.

Early work on visual servoing of wheeled vehicles include those of [22] and [11]. In the latter papers a feedback control law stabilizing the vehicle posture by using visual information only was solved. More recently, the problem under the

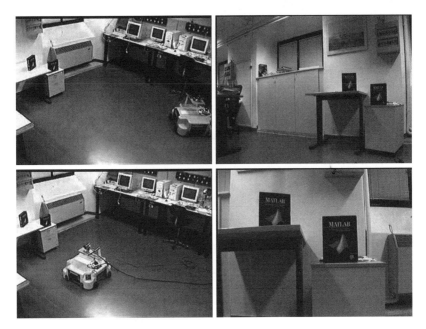

Fig. 8. External views (left column) and subjective images (right column) as taken from the vehicle, in the initial configuration (top row) and in the final configuration (bottom row), after reaching convergence under the visual feedback control scheme of [34]. The bottom right image should be compared with the target image in Fig. 7.

practically most relevant constraint of keeping tracked features within sight of a limited–aperture camera while the vehicle maneuvers to park has been considered in [34]. The method proposed in [34] adopts a hybrid control law, that solves the problem by switching among different stabilizing output-feedback laws, depending on conditions triggered by events such as the approach of image boundaries by some tracked features. It is to be noted that, although different sensors (such as some models of laser range finders, or omnidirectional cameras, or pan-tilt heads) may not be affected by view-angle limitations, these are typically some orders of magnitude more expensive than the conventional cameras considered in [34], which are readily available even in the consumer market. Implementation of the visual-servo method of [34] is based on selecting a few target features from an image recorded at the desired configuration, and by comparing their position in the image plane with that obtained in real time from the robot camera. Some experimental results obtained by application of this method are reported in Figs. 7 and 8.

6   Conclusion

In this chapter, we have considered the connection of three different problems, lo-calization, map building, and servoing, of mobile vehicles moving in unstructured or partially structured environments. An effort has been paid at casting the three problems within a unique framework, which is that provided by the theory of dynamical control systems with outputs. Although this approach is still to be validated in large-scale applications, where the dimensionality of the space of unknowns and the possible topological complexity of the environment can place formidable obstacles, there seems to be some interesting avenue of development at the confluence of classical computer-science and probabilistic approaches and system theory.

Acknowledgement

This work has been co-funded by contracts IST-2001-37170 (RECSYS) and ASI I/R/124/02 (TEMA).

References

1. A. Bicchi and G. Canepa, "Optimal design of multivariate sensors," *Measurement Science and Technology*, vol. 5, pp. 319–332, 1994.
2. A. Bicchi, D. Prattichizzo, A. Marigo, and A. Balestrino, "On the observability of mobile vehicles localization," *Proc. of 6th IEEE Mediterranean Conference on Control and Automation*, pp. 100–105, 1998.
3. G. Bonard, F. Celle-Couenne, and G. Gilles, *Nonlinear Systems, Vol. 1*, Chapman & Hall, 1995.
4. G. Bonard and H. Hammouri, "A high gain observer for a class of uniformly observable systems," *Proc. of 30th IEEE Conference on Decision and Control*, pp. 1494–1496, 1991.
5. J. Borenstein, A. Everett, and L. Feng, *Navigating Mobile Robots*, AK Peters, 1996.
6. G. Campion, B. d'Andréa-Novel, and G. Bastin, "Controllability and state feedback stabilization of nonholonomic wheeled mechanical systems," in C. Canudas de Wit (Ed.), *Advanced Robotic Control*, pp. 106–124, Springer Verlag, 1991.
7. C. Canudas de Wit, and O.J. Sørdalen, "Exponential stabilization of mobile robots with nonholonomic constraints," *IEEE Trans. on Robotics and Automation*, vol. 37, pp. 1791–1797, 1992.
8. G. Casalino, M. Aicardi, A. Bicchi, and A. Balestrino, "Closed loop steering and path following for unicycle-like vehicles: A simple Lyapunov function based approach," *IEEE Robotics and Automation Mag.*, vol. 2, no. 1, pp. 27–35, 1995.
9. F. Celle, J.P. Gauthier, and D. Kazakos, "Orthogonal representations of nonlinear systems and input-output maps," *Systems & Control Letters*, vol. 7, pp. 365–372, 1986.
10. F. Conticelli and A. Bicchi, "Observer design for locally observable analytic systems: Convergence, separation property, and redundancy," in A. Isidori, F. Lamnabhi-Lagarrigue, and W. Respondek (Eds.), *Nonlinear Control in the Year 2000*, pp. 315–330, Springer Verlag, 2000.
11. F. Conticelli, D. Prattichizzo, F. Guidi, and A. Bicchi, "Vision-based dynamic estimation and set-point stabilization of nonholonomic vehicles," *Proc. of 2000 IEEE Int. Conf. on Robotics and Automation*, pp. 2771–2776.
12. B. d'Andréa-Novel, G. Campion, and G. Bastin, "Control of nonholonomic wheeled mobile robots by state feedback linearization," *Int. J. of Robotics Research*, vol. 14, pp. 543–559, 1995.

13. A. De Luca, G. Oriolo, and M. Vendittelli, "Control of wheeled mobile robots: An experimental overview," in S. Nicosia, B. Siciliano, A. Bicchi, and P. Valigi (Eds.), *RAMSETE — Articulated and Mobile Robots for SErvices and TEchnology*, pp. 181–226, Springer Verlag, 2001.

14. A.P. Dempster, A.N. Laird, and D.B. Rubin, "Maximum likelihood from incomplete data via the EM algorithm," *J. of the Royal Statistical Society, Series B*, vol. 39, pp. 1–38,1977.

15. J.W. Fenwick, P.M. Newman, and J.J. Leonard "Cooperative concurrent mapping and localization," *Proc. of 2002 IEEE Int. Conf. on Robotics and Automation*, pp. 1802–1809, 2002.

16. M. Fliess and I. Kupka, "A finiteness criterion for nonlinear input-output differential systems," *SIAM J. of Control and Optimization*, vol. 21, pp. 712–729, 1983.

17. F. Gambino, G. Ulivi, and M. Vendittelli, "The transferable belief model in ultrasonic map building," *Proc. of 6th IEEE Conf. on Fuzzy Systems*, pp. 601–606, 1997.

18. A. Garulli and A. Vicino, "Set membership localization of mobile robots via angle measurements," *IEEE Trans. on Robotics and Automation*, vol. 17, pp. 450–463, 2001.

19. A. Gelb, *Applied Optimal Estimation*, MIT Press, 1974.

20. J.E. Guivant and E.M. Nebot, "Optimization of the simultaneous localization and map-building algorithm for real-time implementation," *IEEE Trans. on Robotics and Automation*, vol. 17, pp. 242–257, 2001.

21. R. Hermann and A. Krener, "Nonlinear controllability and observability," *IEEE Trans. on Automatic Control*, vol. 22, pp. 728–740, 1977.

22. K. Hashimoto and T. Noritsugu, "Visual servoing of nonholonomic cart," *Proc. of 1997 IEEE Int. Conf. on Robotics and Automation*, pp. 1719–1724, 1997.

23. S.A. Hutchinson, G.D. Hager, and P.I. Corke, "A tutorial on visual servo control," *IEEE Trans. on Robotics and Automation*, vol. 5, pp. 651–670, 1996.

24. A. Isidori, *Nonlinear Control Systems*, 3rd Ed., Springer Verlag, 1995.

25. L. Jetto, S. Longhi, and D. Vitali, "Localization of a wheeled mobile robot by sensor data fusion based on a fuzzy logic adapted Kalman filter," *Control Engineering Practice*, vol. 7, pp. 763–771, 1999.

26. R.L. Kosut, A. Arbel, and K.M. Kessler: "Optimal sensor design for state reconstruction," *IEEE Trans. on Automatic Control*, vol. 27, pp. 242-244, 1982.

27. A.J. Krener and H. Schättler, "The structure of small time reachable sets in low dimensions," *SIAM J. of Control and Optimization*, vol. 27, pp. 120–147, 1989.

28. J.J. Leonard, H.F. Durrant-Whyte, and I.J. Cox, "Dynamic map building for an autonomous mobile robot," *Int. J. of Robotics Research*, vol. 11, pp. 89–96, 1992.

29. J. Levine and R. Marino, "Nonlinear system immersion, observers and finite dimensional filters," *Systems & Control Letters*, vol. 7, pp. 137–142, 1986.

30. F. Lorussi, A. Marigo, and A. Bicchi, "Optimal exploratory paths for a mobile rover," *Proc. of 2001 IEEE Int. Conf. on Robotics and Automation*, pp. 2078–2083, 2001.

31. E.A. Misawa and J.K. Hedrick, "Nonlinear observers – A state of the art survey," *ASME J. of Dynamic Systems, Measurement, and Control*, vol. 111, pp. 344–352, 1989.

32. M. Montemerlo, S. Thrun, D. Koller, and B. Wegbreit, "A factored solution to the simultaneous localization and mapping problem," *Proc. of AAAI Nat. Conf. on Artificial Intelligence*, pp. 375–380, 2002.

33. P. Morin and C. Samson, *Application of Backstepping Techniques to the Time-Varying Exponential Stabilization of Chained Form Systems*, INRIA Research Report, Sophia-Antipolis, no. 2792, 1996.

34. P. Murrieri, D. Fontanelli, and A. Bicchi, "Visual-servoed parking with limited view angle," in B. Siciliano and P. Dario (Eds.), *Experimental Robotics VIII*, pp. 254–263, Springer Verlag, 2002.

35. S. Pancanti, L. Leonardi, L. Pallottino, and A. Bicchi, "Optimal control of quantized input systems," in C. Tomlin and M. Greenstreet (Eds.), *Hybrid Systems: Computation and Control*, pp. 351–363, Springer Verlag, 2002.

36. F. Piloni and A. Bicchi, *Navigazione di Veicoli Autonomi Mediante Visione Artificiale: Pianificazione dei Moti di Esplorazione*, Internal Report, Centro Piaggio, University of Pisa, 1994.

37. C.R. Rao, *Linear Statistical Inference and Its Applications*, Wiley, 1973.

38. S. Se, D. Lowe, and J. Little, "Mobile robot localization and mapping with uncertainty using scale-invariant visual landmarks," *Int. J. of Robotics Research*, vol. 21, pp. 735–758, 2002.

39. A.R. Teel and L. Praly, "Global stabilizability and observability imply semi-global stabilizability by output feedback," *Systems & Control Letters*, vol. 22, pp. 313–325, 1994.

40. S. Thrun, "Robotic mapping: A survey," in G. Lakemeyer and B. Nebel (Eds.), *Exploring Artificial Intelligence in the New Millenium*, Morgan Kaufmann, 2002.

41. S. Thrun, W. Burgard, and D. Fox, "A probabilistic approach to concurrent mapping and localization for mobile robots," *Machine Learning and Autonomous Robots*, vol. 31, pp. 1-25, 1998.

42. S. Thrun, D. Koller, Z. Ghahmarani, and H. Durrant-Whyte, *SLAM Updates Require Constant Time*, Tech. rep., School of Computer Science, Carnegie Mellon University, 2002.

43. A. Tarantola, *Inverse Problem Theory*, Elsevier, 1987.

44. D. Tilbury and A. Chelouah, "Steering a three-input nonholonomic system using multi-rate controls," *Proc. of 2nd European Control Conf.*, pp. 1428–1431, 1993.

45. M. Zeitz, "The extended Luenberger observer for nonlinear systems," *Systems & Control Letters*, vol. 9, pp. 149–156, 1987.

Springer Tracts in Advanced Robotics

Edited by B. Siciliano, O. Khatib, and F. Groen
Published Titles:

Printing: Saladruck, Berlin
Binding: Stein+Lehmann, Berlin

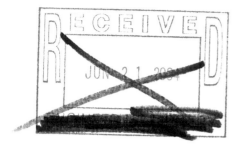